THE GOLDEN [illegible]

This riveting n[illegible] adventure is based on a true story, one of the most daring hijacking exploits in recent history, which continues to baffle both the Italian police and Interpol.

When the Allies landed in Italy during the last war, Mussolini's vast personal treasure, consisting of four tons of gold, millions in currency and jewels, and some of the most important Government archives, was moved north in a German S.S. convoy. As the convoy neared the Ligurian coast, it vanished. It has never been recovered.

But a few men knew where it was hidden. Years later they decided to try and smuggle it out of Italy. For this purpose they built a specially designed yacht in South Africa and set sail for the Mediterranean. But getting the treasure out of Italy proved well-nigh impossible. Not only were the Italian Government hot on the trail, but also a group of former partisans as well as a piratical British smuggler. The fate of the yacht and her crew is charted with breathtaking skill in a tremendous sea chase across the Mediterranean.

Available in Fontana by the same author

Running Blind
Wyatt's Hurricane
Landslide
The Vivero Letter
High Citadel
The Spoilers

DESMOND BAGLEY

The Golden Keel

FONTANA / Collins

First published 1963
First issued in Fontana Books 1965
Fourteenth Impression November 1973

Printed in Great Britain
Collins Clear-Type Press
London and Glasgow

CONTENTS

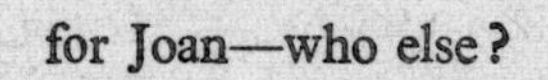

for Joan—who else?

BOOK ONE: THE MEN

1. *WALKER*

My name is Peter Halloran, but everyone calls me "Hal" excepting my wife, Jean, who always called me Peter. Women seem to dislike nicknames for their menfolk. Like a lot of others I emigrated to the "colonies" after the war, and I travelled from England to South Africa by road, across the Sahara and through the Congo. It was a pretty rough trip, but that's another story; it's enough to say that I arrived in Cape Town in 1948 with no job and precious little money.

During my first week in Cape Town I answered several of the Sit. Vac. advertisements which appeared in the *Cape Times* and while waiting for answers I explored my environment. On this particular morning I had visited the docks and finally found myself near the yacht basin.

I was leaning over the rail looking at the boats when a voice behind me said, "If you had your choice, which would it be?"

I turned and encountered the twinkling eyes of an elderly man, tall, with stooped shoulders and grey hair. He had a brown, weather-beaten face and gnarled hands, and I estimated his age at about sixty.

I pointed to one of the boats. "I think I'd pick that one," I said. "She's big enough to be of use, but not too big for single-handed sailing."

He seemed pleased. "That's *Gracia,*" he said. "I built her."

"She looks a good boat," I said. "She's got nice lines."

We talked for a while about boats. He said that he had a boatyard a little way outside Cape Town towards Milnerton, and that he specialised in building the fishing boats used by the Malay fishermen. I'd noticed these already; sturdy unlovely craft with high bows and a wheelhouse stuck on top like a chicken-coop, but they looked very seaworthy. *Gracia* was only the second yacht he had built.

"There'll be a boom now the war's over," he predicted. "People will have money in their pockets, and they'll go in for yachting. I'd like to expand my activities in that direction."

Presently he looked at his watch and nodded towards the yacht club. "Let's go in and have a coffee," he suggested.

I hesitated. " I'm not a member."

" I am," he said. " Be my guest."

So we went into the club house and sat in the lounge overlooking the yacht basin and he ordered coffee. By the way, my name's Tom Sanford."

" I'm Peter Halloran."

" You're English," he said. " Been out here long?"

I smiled. " Three days."

" I've been out just a bit longer—since 1910." He sipped his coffee and regarded me thoughtfully. " You seem to know a bit about boats."

" I've been around them all my life," I said. " My father had a boatyard on the east coast, quite close to Hull. We built fishing boats, too, until the war."

" And then?"

" Then the yard went on to contract work for the Admiralty," I said. " We built harbour defence launches and things like that—we weren't geared to handle anything bigger." I shrugged. " Then there was an air-raid."

" That's bad," said Tom. " Was everything destroyed?"

" Everything," I said flatly. " My people had a house next to the yard—that went, too. My parents and my elder brother were killed."

" Christ!" said Tom gently. " That's very bad. How old were you?"

" Seventeen," I said. " I went to live with an aunt in Hatfield; that's when I started to work for de Havilland—building Mosquitos. It's a wooden aeroplane and they wanted people who could work in wood. All I was doing, as far as I was concerned, was filling in time until I could join the Army."

His interest sharpened. " You know, that's the coming thing—the new methods developed by de Havilland. That hot-moulding process of theirs—d'you think it could be used in boat-building?"

I thought about it. " I don't see why not—it's very strong. We did repair work at Hatfield, as well as new construction, and I saw what happens to that type of fabric when it's been hit very hard. It would be more expensive than the traditional methods, though, unless you were mass-producing."

" I was thinking about yachts," said Tom slowly. " You must tell me more about it sometime." He smiled. " What else do you know about boats?"

I grinned. " I once thought I'd like to be a designer," I said.

" When I was a kid—about fifteen—I designed and built my own racing dinghy."

" Win any races?"

" My brother and I had 'em all licked," I said. " She was a fast boat. After the war, when I was cooling my heels waiting for my discharge, I had another go at it—designing, I mean. I designed half a dozen boats—it helped to pass the time."

" Got the drawings with you?"

" They're somewhere at the bottom of my trunk," I said. " I haven't looked at them for a long time."

" I'd like to see them," said Tom. " Look, laddie; how would you like to work for me? I told you I'm thinking of expanding into the yacht business, and I could use a smart young fellow."

And that's how I started working for Tom Sanford. The following day I went to the boatyard with my drawings and showed them to Tom. On the whole he liked them, but pointed out several ways in which economies could be made in the building. " You're a fair designer," he said. " But you've a lot to learn about the practical side. Never mind, we'll see about that. When can you start?"

Going to work for old Tom was one of the best things I ever did in my life.

II

A lot of things happened in the next ten years—whether I deserved them or not is another matter. It was good to work in a boatyard again. The skills I had learned from my father had not deserted me, and although I was a bit rusty to begin with, soon I was as good as any man in the yard, and maybe a bit better. Tom encouraged me to design, ruthlessly correcting my errors.

" You've got a good eye for line," he said. " Your boats would be sweet sailers, but they'd be damned expensive. You've got to spend more time on detail; you must cut down costs to make an economical boat."

Four years after I joined the firm Tom made me yard foreman, and just after that, I had my first bit of luck in designing. I submitted a design to a local yachting magazine, winning second prize and fifty pounds. But better still, a local yachtsman liked the design and wanted a boat built. So Tom built it for him and I got the designers' fee which went to swell my growing bank balance.

Tom was pleased about that and asked if I could design a class boat as a standard line for the yard, so I designed a six-tonner which turned out very well. We called it the Penguin Class and Tom built and sold a dozen in the first year at £2000 each. I liked the boat so much that I asked Tom if he would build one for me, which he did, charging a rock-bottom price and letting me pay it off over a couple of years.

Having a design office gave the business a fillip. The news got around and people started to come to me instead of using British and American designs. That way they could argue with their designer. Tom was pleased because most of the boats to my design were built in the yard.

In 1954 he made me yard manager, and in 1955 he offered me a partnership.

"I've got no one to leave it to," he said bluntly. "My wife's dead and I've got no sons. And I'm getting old."

I said, "You'll be building boats when you're a hundred, Tom."

He shook his head. "No, I'm beginning to feel it now." He wrinkled his brow. "I've been going over the books and I find that you're bringing more business into the firm than I am, so I'll go easy on the money for the partnership. It'll cost you five thousand pounds."

Five thousand was ridiculously cheap for a half-share in such a flourishing business, but I hadn't got anywhere near that amount. He saw my expression and his eyes crinkled. "I know you haven't got it—but you've been doing pretty well on the design side lately. My guess is that you've got about two thousand salted away."

Tom, shrewd as always, was right. I had a couple of hundred over the two thousand. "That's about it," I said.

"All right. Throw in the two thousand and borrow another three from the bank. They'll lend it to you when they see the books. You'll be able to pay it back out of profits in under three years, especially if you carry out your plans for that racing dinghy. What about it?"

"O.K., Tom," I said. "It's a deal."

The racing dinghy Tom had mentioned was an idea I had got by watching the do-it-yourself developments in England. There are plenty of little lakes on the South African high-veld and I thought I could sell small boats away from the sea if I could produce them cheaply enough—and I would sell either the finished boat or a do-it-yourself building kit for the impoverished enthusiast.

We set up another woodworking shop and I designed the

boat which was the first of the Falcon Class. A young fellow, Harry Marshall, was promoted to run the project and he did very well. This wasn't Tom's cup of tea and he stayed clear of the whole affair, referring to it as "that confounded factory of yours." But it made us a lot of money.

It was about this time that I met Jean and we got married. My marriage to Jean is not really a part of this story and I wouldn't mention it except for what happened later. We were very happy and very much in love. The business was doing well—I had a wife and a home—what more could a man wish for?

Towards the end of 1956 Tom died quite suddenly of a heart attack. I think he must have known that his heart wasn't in good shape although he didn't mention it to anyone. He left his share in the business to his wife's sister. She knew nothing about business and less about boat-building, so we got the lawyers on to it and she agreed to sell me her share. I paid a damn' sight more than the five thousand I had paid Tom, but it was a fair sale although it gave me financier's fright and left me heavily in debt to the bank.

I was sorry that Tom had gone. He had given me a chance that fell to few young fellows and I felt grateful. The yard seemed emptier without him pottering about the slips.

The yard prospered and it seemed that my reputation as a designer was firm, because I got lots of commissions. Jean took over the management of the office, and as I was tied to the drawing board for a large proportion of my time I promoted Harry Marshall to yard manager and he handled it very capably.

Jean, being a woman, gave the office a thorough spring cleaning as soon as she was in command, and one day she unearthed an old tin box which had stayed forgotten on a remote shelf for years. She delved into it, then said suddenly, "Why have you kept this clipping?"

"What clipping?" I asked abstractedly. I was reading a letter which could lead to an interesting commission.

"This thing about Mussolini," she said. "I'll read it." She sat on the edge of the desk, the yellowed fragment of newsprint between her fingers. "'Sixteen Italian Communists were sentenced in Milan yesterday for complicity in the disappearance of Mussolini's treasure. The treasure, which mysteriously vanished at the end of the war, consisted of a consignment of gold from the Italian State Bank and many of Mussolini's personal possessions, including the Ethiopian crown. It is believed that a large number of important State documents were with the treasure. The sixteen men all declared their innocence.' "

She looked up. " What was all that about."

I was startled. It was a long time since I'd thought of Walker and Coertze and the drama that had been played out in Italy. I smiled and said, " I might have made a fortune but for that news story."

" It's a long story," I protested. " I'll tell you some other time."

" No," she insisted. " Tell me now; I'm always interested in treasure."

So I pushed the unopened mail aside and told her about Walker and his mad scheme. It came back to me hazily in bits and pieces. Was it Donato or Alberto who had fallen—or been pushed—from the cliff? The story took a long time in the telling and the office work got badly behind that day.

III

I met Walker when I had arrived in South Africa from England just after the war. I had been lucky to get a good job with Tom but, being a stranger, I was a bit lonely, so I joined a Cape Town Sporting Club which would provide company and exercise.

Walker was a drinking member, one of those crafty people who joined the club to have somewhere to drink when the pubs were closed on Sunday. He was never in the club house during the week, but turned up every Sunday, played his one game of tennis for the sake of appearances, then spent the rest of the day in the bar.

It was in the bar that I met him, late one Sunday afternoon. The room was loud with voices raised in argument and I soon realised I had walked into the middle of a discussion on the Tobruk surrender. The very mention of Tobruk can start an argument anywhere in South Africa because the surrender is regarded as a national disgrace. It is always agreed that the South Africans were let down but from then on it gets heated and rather vague. Sometimes the British generals are blamed and sometimes the South African garrison commander, General Klopper; and it's always good for one of those long, futile bar-room brawls in which tempers are lost but nothing is ever decided.

It wasn't of much interest to me—my army service was in Europe—so I sat quietly nursing my beer and keeping out of it. Next to me was a thin-faced young man with dissipated good looks who had a great deal to say about it, with many a thump on the counter with his clenched fist. I had seen him before but didn't know who he was. All I knew of him was by observation; he seemed to drink a lot, and even now was drinking two brandies to my one beer.

At length the argument died a natural death as the bar emptied and soon my companion and I were the last ones left. I drained my glass and was turning to leave when he said contemptuously, "Fat lot they know about it."

"Were you there?" I asked.

"I was," he said grimly. "I was in the bag with all the others. Didn't stay there long, though; I got out of the camp in Italy in '43." He looked at my empty glass. "Have one for the road."

I had nothing to do just then, so I said, "Thanks; I'll have a beer."

He ordered a beer for me and another brandy for himself and said, "My name's Walker. Yes, I got out when the Italian Government collapsed. I joined the partisans."

"That must have been interesting," I said.

He laughed shortly. "I suppose you could call it that. Interesting and scary. Yes, I reckon you could say that me and Sergeant Coertze had a really interesting time—he was a bloke I was with most of the time."

"An Afrikaner?" I hazarded. I was new in South Africa and didn't know much about the set-up then, but the name sounded as though it might be Afrikaans.

"That's right," said Walker. "A real tough boy, he was. We stuck together after getting out of the camp."

"Was it easy—escaping from the prison camp?"

"A piece of cake," said Walker. "The guards co-operated with us. A couple of them even came with us as guides—Alberto Corso and Donato Rinaldi. I liked Donato—I reckon he saved my life."

He saw my interest and plunged into the story with gusto. When the Government fell in 1943 Italy was in a mess. The Italians were uneasy; they didn't know what was going to happen next and they were suspicious of the intentions of the Germans. It was a perfect opportunity to break camp, especially when a couple of the guards threw in with them.

Leaving the camp was easy enough, but trouble started soon after when the Germans laid on an operation to round up all the Allied prisoners who were loose in Central Italy.

"That's when I copped it," said Walker. "We were crossing a river at the time."

The sudden attack had taken them by surprise. Everything had been silent except for the chuckling of the water and the muffled curses as someone slipped—then suddenly there was the sound of ripped calico as the Spandau opened up and the night was made hideous by the eerie whine of bullets as they ricocheted from exposed rocks in the river.

The two Italians turned and let go with their sub-machine-guns. Coertze, bellowing like a bull, scrabbled frantically at the pouch pocket of his battle-dress trousers and then his arm came up in an overarm throw. There was a sharp crack as the hand grenade exploded in the water near the bank. Again Coertze threw and this time the grenade burst on the bank.

Walker felt something slam his leg and he turned in a twisting fall and found himself gasping in the water. His free arm thrashed out and caught on a rock and he hung on desperately.

Coertze threw another grenade and the machine-gun stopped. The Italians had emptied their magazines and were busy reloading. Everything was quiet again.

"I reckon they thought we were Germans, too," said Walker. "They wouldn't expect to be fired on by escaping prisoners. It was lucky that the Italians had brought some guns along. Anyway, that bloody machine-gun stopped."

They had stayed for a few minutes in midstream with the quick cold waters pulling at their legs, not daring to move in case there was a sudden burst from the shore. After five minutes Alberto said in a low voice, "Signor Walker, are you all right?"

Walker pulled himself upright and to his astonishment found himself still grasping his unfired rifle. His left leg felt numb and cold. "I'm all right," he said.

There was a long sigh from Coertze, then he said, "Well, come on. Let's get to the other side—but quietly."

They reached the other side of the river and, without resting, pressed on up the mountainside. After a short time Walker's leg began to hurt and he lagged behind. Alberto was perturbed. "You must hurry; we have to cross this mountain before dawn."

Walker stifled a groan as he put down his left foot. "I was hit," he said. "I think I was hit."

Coertze came back down the mountain and said irritably, "*Magtig,* get a move on, will you?"

Alberto said, "Is it bad, Signor Walker?"

"What's the matter?" asked Coertze, not understanding the Italian.

"I have a bullet in my leg," said Walker bitterly.

"That's all we need," said Coertze. In the darkness he bulked as a darker patch and Walker could see that he was shaking his head impatiently. "We've got to get to that partisan camp before daylight."

Walker conferred with Alberto, then said in English, "Alberto says there's a place along there to the right where we can hide. He says that someone should stay with me while he goes for help."

Coertze grunted in his throat. "I'll go with him," he said. "The other Eytie can stay with you. Let's get to it."

They moved along the mountainside and presently the ground dipped and suddenly there was a small ravine, a cleft in the mountain. There were stunted trees to give a little cover and underfoot was a dry watercourse.

Alberto stopped and said, "You will stay here until we come for you. Keep under the trees so that no one will see you, and make as little movement as possible."

"Thanks, Alberto," said Walker. There were a few brief words of farewell, then Alberto and Coertze disappeared into the night. Donato made Walker comfortable and they settled down to wait out the night.

It was a bad time for Walker. His leg was hurting and it was very cold. They stayed in the ravine all the next day and as night fell Walker became delirious and Donato had trouble in keeping him quiet.

When the rescuers finally came Walker had passed out. He woke up much later and found himself in a bed in a room with whitewashed walls. The sun was rising and a little girl was sitting by the bedside.

Walker stopped speaking suddenly and looked at his empty glass on the bar counter. "Have another drink," I said quickly.

He needed no encouraging so I ordered another couple of drinks. "So that's how you got away," I said.

He nodded. "That's how it was. God, it was cold those two nights on that bloody mountain. If it hadn't been for Donato I'd have cashed in my chips."

I said, "So you were safe—but where were you?"

"In a partisan camp up in the hills. The *partigiani* were just getting organised then; they only really got going when the Germans began to consolidate their hold on Italy. The Jerries ran true to form—they're arrogant bastards, you know—and the Italians didn't like it. So everything was set for the partisans; they got the support of the people and they could begin to operate on a really large scale.

"They weren't all alike, of course; there was every shade of political opinion from pale blue to bright red. The Communists hated the Monarchists' guts and vice versa and so on. The crowd I dropped in on were Monarchist. That's where I met the Count."

Count Ugo Montepescali di Todi was over fifty years old at that time, but young-looking and energetic. He was a swarthy man with an aquiline nose and a short greying beard which was split at the end and forked aggressively. He came of a line which was old during the Renaissance and he was an aristocrat to his fingertips.

Because of this he hated Fascism—hated the pretensions of the parvenu rulers of Italy with all their corrupt ways and their money-sticky fingers. To him Mussolini always remained a mediocre journalist who had succeeded in demagoguery and had practically imprisoned his King.

Walker met the Count the first day he arrived at the hill camp. He had just woken up and seen the solemn face of the little girl. She smiled at him and silently left the room, and a few minutes later a short stocky man with a bristling beard stepped through the doorway and said in English, "Ah, you are awake. You are quite safe now."

Walker was conscious of saying something inane. "But where am I?"

"Does that really matter?" the Count asked quizzically. "You are still in Italy—but safe from the *Tedesci*. You must stay in bed until you recover your strength. You need some blood putting back—you lost a lot—so you must rest and eat and rest again."

Walker was too weak to do more than accept this, so he lay back on the pillow. Five minutes later Coertze came in; with him was a young man with a thin face.

"I've brought the quack," said Coertze. "Or at least that's what he says he is—if I've got it straight. My guess is that he's only a medical student."

The doctor—or student—examined Walker and professed satisfaction at his condition. "You will walk within the week," he said, and packed his little kit and left the room.

Coertze rubbed the back of his head. "I'll have to learn this slippery *taal*," he said. "It looks as though we'll be here for a long time."

"No chance of getting through to the south?" asked Walker.

"No chance at all," said Coertze flatly. "The Count—that's the little man with the *bokbaardjie*—says that the Germans down south are thicker on the ground than stalks in a mealie field. He reckons they're going to make a defence line south of Rome."

Walker sighed. "Then we're stuck here."

Coertze grinned. "It is not too bad. At least we'll get better food than we had in camp. The Count wants us to join his little lot—it seems he has some kind of *skietkommando* which holds quite a bit of territory and he's collected men and weapons while he can. We might as well fight here as with the army—I've always fancied fighting a war my way."

A plump woman brought in a steaming bowl of broth for Walker, and Coertze said, "Get outside of that and you'll feel better. I'm going to scout around a bit."

Walker ate the broth and slept, then woke and ate again. After a while a small figure came in bearing a basin and rolled bandages. It was the little girl he had seen when he had first opened his eyes. He thought she was about twelve years old.

"My father said I had to change your bandages," she said in a clear young voice. She spoke in English.

Walker propped himself up on his elbows and watched her as she came closer. She was neatly dressed and wore a white, starched apron. "Thank you," he said.

She bent to cut the splint loose from his leg and then she carefully loosened the bandage round the wound. He looked down at her and said, "What is your name?"

"Francesca."

"Is your father the doctor?" Her hands were cool and soft on his leg.

She shook her head. "No," she said briefly.

She bathed the wound in warm water containing some pungent antiseptic and then shook powder on to it. With great skill she began to rebandage the leg.

"You are a good nurse," said Walker.

It was only then that she looked at him and he saw that she had cool, grey eyes. "I've had a lot of practice," she said, and Walker was abashed at her gaze and cursed a war which made skilled nurses out of twelve-year-olds.

She finished the bandaging and said, "There—you must get better soon."

"I will," promised Walker. "As quickly as I can. I'll do that for you."

She looked at him with surprise. "Not for me," she said. "For the war. You must get better so that you can go into the hills and kill a lot of Germans."

She gravely collected the soiled bandages and left the room, with Walker looking after her in astonishment. Thus it was that he met Francesca, the daughter of Count Ugo Montepescali.

In a little over a week he was able to walk with the aid of a stick and to move outside the hospital hut, and Coertze showed him round the camp. Most of the men were Italians, army deserters who didn't like the Germans. But there were many Allied escapees of different nationalities.

The Count had formed the escapees into a single unit and had put Coertze in command. They called themselves the "Foreign Legion." During the next couple of years many of them were to be killed fighting against the Germans with the partisans. At Coertze's request, Alberto and Donato were attached to the unit to act as interpreters and guides.

Coertze had a high opinion of the Count. "That *kêrel* knows what he's doing," he said. "He's recruiting from the Italian army as fast as he can—and each man must bring his own gun."

When the Germans decided to stand and fortified the *Winterstellung* based on the Sangro and Monte Cassino, the war in Italy was deadlocked and it was then that the partisans got busy attacking the German communications. The Foreign Legion took part in this campaign, specialising in demolition work. Coertze had been a gold miner on the Witwatersrand before the war and knew how to handle dynamite. He and Harrison, a Canadian geologist, instructed the others in the use of explosives.

They blew up road and rail bridges, dynamited mountain passes, derailed trains and occasionally shot up the odd road convoy, always retreating as soon as heavy fire was returned. "We must not fight pitched battles," said the Count. "We must not let the Germans pin us down. We are mosquitoes irritating the German hides—let us hope we give them malaria."

Walker found this a time of long stretches of relaxation punctuated by moments of fright. Discipline was easy and there was no army spit-and-polish. He became lean and hard and would think nothing of making a day's march of thirty miles over the mountains burdened with his weapons and a pack of dynamite and detonators.

By the end of 1944 the Foreign Legion had thinned down considerably. Some of the men had been killed and more elected to make a break for the south after the Allies had taken Rome. Coertze said he would stay, so Walker stayed with him. Harrison also stayed, together with an Englishman called Parker. The Foreign Legion was now very small indeed.

"The Count used us as bloody pack horses," said Walker. He had ordered another round of drinks and the brandy was getting at him. His eyes were red-veined and he stumbled over the odd word.

"Pack horses?" I queried.

"The unit was too small to really fight," he explained. "So he used us to transport guns and food around his territory. That's how we got the convoy."

"Which convoy?"

Walker was beginning to slur his words. "It was like this. One of the Italian units had gone to carve up a German post and the job was being done in co-operation with another partisan brigade. But the Count was worried because this other mob were Communists—real treacherous bastards they were. He was scared they might renege on us; they were always doing that because he was a Monarchist and they hated him worse than they did the Germans. They were looking ahead to after the war and they didn't do much fighting while they were about it. Italian politics, you see."

I nodded.

"So he wanted Umberto—the chap in charge of our Italians—to have another couple of machine-guns, just in case, and Coertze said he'd take them."

He fell silent, looking into his glass.

I said, "What about this convoy?"

"Oh, what the hell," he said. "There's not a hope of getting it out. It'll stay there for ever, unless Coertze does something. I'll tell you. We were on our way to Umberto when we bumped into this German convoy driving along where no convoy should have been. So we clobbered it."

They had got to the top of a hill and Coertze called a halt. "We stay here for ten minutes, then we move on," he said.

Alberto drank some water and then strolled down to where he could get a good view of the valley. He looked first at the valley floor where a rough, unmetalled road ran dustily, then raised his eyes to look south.

Suddenly he called Coertze. "Look," he said.

Coertze ran down and looked to where Alberto was pointing. In the distance, where the faraway thread of brown road shimmered in the heat, was a puff of dust. He unslung his glasses and focused rapidly.

"What the hell are they doing here?" he demanded. "What is it?"

"German army trucks," said Coertze. "About six of them." He pulled down the glasses. "Looks as though they're trying to slip by on the side roads. We *have* made the main roads a bit unhealthy."

Walker and Donato had come down. Coertze looked back at the machine-guns, then at Walker. "What about it?"

Walker said, "What about Umberto?"

"Oh, he's all right. It's just the Count getting a bit fretful now the war's nearly over. I think we should take this little lot—it should be easy with two machine-guns."

Walker shrugged. "O.K. with me," he said.

Coertze said, "Come on," and ran back to where Parker was sitting. "On your feet, *kêrel*," he said. "The war's still on. Where the hell is Harrison?"

"Coming," called Harrison.

"Let's get this stuff down to the road on the double," said Coertze. He looked down the hill. "That bend ought to be a *lekker* place."

"A what?" asked Parker plaintively. He always pulled Coertze's leg about his South Africanisms.

"Never mind that," snapped Coertze. "Get this stuff down to the road quick. We've got a job on."

They loaded up the machine-guns and plunged down the hillside. Once on the road Coertze did a quick survey. "They'll come round that bend slowly," he said. "Alberto, you take Donato and put your machine-gun there, where you can open up on the last two trucks. The last two, you understand. Knock 'em out fast so the others can't back out."

He turned to Harrison and Parker. "Put you gun over here on the other side and knock out the first truck. Then we'll have the others boxed in."

"What do I do?" asked Walker.

"You come with me." Coertze started to run up the road, followed by Walker. He ran almost to the bend, then left the road and climbed a small hillock from where he could get a good sight of the German convoy. When Walker flopped beside him he already had the glasses focused.

"It's four trucks not six," he said. "There's a staff car in front and a motor-cycle combination in front of that. Looks like one of those BMW jobs with a machine-gun in the side-car."

He handed the glasses to Walker. "How far from the tail of the column to that staff car?"

Walker looked at the oncoming vehicles. "About sixty-five yards," he estimated.

Coertze took the glasses. "O.K. You go back along the road sixty-five yards so that when the last truck is round the bend the staff car is alongside you. Never mind the motor-cycle—I'll take care of that. Go back and tell the boys not to open up until they hear loud bangs; I'll start those off. And tell them to concentrate on the trucks."

He turned over and looked back. The machine-guns were invisible and the road was deserted. "As nice an ambush as anyone could set," he said. "My *oupa* never did better against the English." He tapped Walker on the shoulder. "Off you go. I'll help you with the staff car as soon as I've clobbered the motor-cycle."

Walker slipped from the hillock and ran back along the road, stopping at the machine-guns to issue Coertze's instructions. Then he found himself a convenient rock about sixty yards from the bend, behind which he crouched and checked his sub-machine-gun.

It was not long before he heard Coertze running along the road shouting, "Four minutes. They'll be here in four minutes. Hold your fire."

Coertze ran past him and disappeared into the verge of the road about ten yards farther on.

Walker said that four minutes in those conditions could seem like four hours. He crouched there, looking back along the silent road, hearing nothing except his own heart beating. After what seemed a long time he heard the growl of engines and the clash of gears and then the revving of the motor-cycle.

He flattened himself closer to the rock and waited. A muscle twitched in his leg and his mouth was suddenly dry. The noise of the motor-cycle now blanked out all other sounds and he snapped off the safety catch.

He saw the motor-cycle pass, the goggled driver looking like a gargoyle and the trooper in the sidecar turning his head to scan the road, hands clutching the grips of the machine-gun mounted in front of him.

As in a dream he saw Coertze's hand come into view, apparently in slow motion, and toss a grenade casually into the sidecar. It lodged between the gunner's back and the coaming of the sidecar and the gunner turned in surprise. With his sudden movement the grenade disappeared into the interior of the sidecar.

Then it exploded.

The sidecar disintegrated and the gunner must have had his

legs blown off. The cycle wheeled drunkenly across the road and Walker saw Coertze step out of cover, his sub-machine-gun pumping bullets into the driver. Then he had stepped out himself and his own gun was blazing at the staff car.

He had orientated himself very carefully so that he had a very good idea of where the driver would be placed. When he started firing, he did so without aiming and the windscreen shattered in the driver's face.

In the background he was conscious of the tac-a-tac of the machine-guns firing in long bursts at the trucks, but he had no time or desire to cast a glance that way. He was occupied in jumping out of the way of the staff car as it slewed towards him, a dead man's hand on the wheel.

The officer in the passenger seat was standing up, his hand clawing at the flap of his pistol holster. Coertze fired a burst at him and he suddenly collapsed and folded grotesquely over the metal rim of the broken windscreen as though he had suddenly turned into a rag doll. The pistol dropped from hand and clattered on the ground.

With a rending jar the staff car bumped into a rock on the side of the road and came to a sudden stop, jolting the soldier in the rear who was shooting at Walker. Walker heard the bullets going over his head and pulled the trigger. A dozen bullets hit the German and slammed him back in his seat. Walker said that the range was about nine feet and he swore he heard the bullets hit, sounding like a rod hitting a soft carpet several times.

Then Coertze was shouting at him, waving him on to the trucks. He ran up the road following Coertze and saw that the first truck was stopped. He fired a burst into the cab just to be on the safe side, then took shelter, leaning against the hot radiator to reload.

By the time he had reloaded the battle was over. All the vehicles were stopped and Alberto and Donato were escorting a couple of dazed prisoners forward.

Coertze barked, "Parker, go up and see if anyone else is coming," then turned to look at the chaos he had planned.

The two men with the motor-cycle had been killed outright, as had the three in the staff car. Each truck had carried two men in the cab and one in the back. All the men in the cabs had been killed within twenty seconds of the machine-guns opening fire. As Harrison said, "At twenty yards we couldn't miss—we just squirted at the first truck, then hosed down the second. It was like using a howitzer at a coconut-shy—too easy."

Of the seventeen men in the German party there were two survivors, one of whom had a flesh wound in his arm.

Coertze said, "Notice anything?"

Walker shook his head. He was trembling in the aftermath of danger and was in no condition to be observant.

Coertze went up to one of the prisoners and fingered the emblem on his collar. The man cringed.

"These are S.S. men. All of them."

He turned and went back to the staff car. The officer was lying on his back, half in and half out the front door, his empty eyes looking up at the sky, terrible in death. Coertze looked at him, then leaned over and pulled a leather brief-case from the front seat. It was locked.

"There's something funny here," he said. "Why would they come by this road?"

Harrison said, "They might have got through, you know. If we hadn't been here they would have got through—and we were only here by chance."

"I know," said Coertze. "They had a good idea and they nearly got away with it—that's what I'm worrying about. The Jerries aren't an imaginative lot, usually; they follow a routine. So why would they do something different? Unless this wasn't a routine unit."

He looked at the trucks. "It might be a good idea to see what's in those trucks."

He sent Donato up the road to the north to keep watch and the rest went to investigate the trucks, excepting Alberto who was guarding the prisoners.

Harrison looked over the tailboard of the first truck. "Not much in here," he said.

Walker looked in and saw that the bottom of the truck was filled with boxes—small wooden boxes about eighteen inches long, a foot wide and six inches deep. He said, "That's a hell of a small load."

Coertze frowned and said, "Boxes like that ring a bell with me, but I just can't place it. Let's have one of them out."

Walker and Harrison climbed into the truck and moved aside the body of a dead German which was in the way. Harrison grasped the corner of the nearest box and lifted. "My God!" he said. "The damn' thing's nailed to the floor."

Walker helped him and the box shifted. "No, it isn't, but it must be full of lead."

Coertze let down the tailboard. "I think we'd better have it out and opened," he said. His voice was suddenly croaking with excitement.

Walker and Harrison manhandled a box to the edge and tipped it over. It fell with a loud thump to the dusty road. Coertze said, "Give me that bayonet."

Walker took the bayonet from the scabbard of the dead German and handed it to Coertze, who began to prise the box open. Nails squealed as the top of the box came up. Coertze ripped it off and said, "I thought so."

"What is it?" asked Harrison, mopping his brow.

"Gold," said Coertze softly.

Everyone stood very still.

Walker was very drunk when he got to this point of his story. He was unsteady on his feet and caught the edge of the bar counter to support himself as he repeated solemnly, "Gold."

"For the love of Mike, what did you do with it?" I said. "And how much of it was there?"

Walker hiccoughed gently. "What about another drink?" he said.

I beckoned to the bar steward, then said, "Come on; you can't leave me in suspense."

He looked at me sideways. "I really shouldn't tell," he said. "But what the hell! There's no harm in it now. It was like this . . ."

They had stood looking at each other for a long moment, then Coertze said, "I knew I recognised those boxes. They use boxes like that on the Reef for packing the ingots for shipment."

As soon as they had checked that all the boxes in that truck were just as heavy, there was a mad rush to the other trucks. These were disappointing at first—the second truck was full of packing cases containing documents and files.

Coertze delved into a case, tossing papers out, and said, "What the hell's all this bumph?" He sounded disappointed.

Walker picked up a sheaf and scanned through it. "Seems to be Italian Government documents of some sort. Maybe this is all top-secret stuff."

The muffled voice of Harrison came from the bowels of the truck. "Hey, you guys, look what I've found."

He emerged with both hands full of bundles of lire notes—fine, newly printed lire notes. "There's at least one case full of this stiff," he said. "Maybe more."

The third truck had more boxes of gold, though not as much as the first, and there were several stoutly built wooden cases which were locked. They soon succumbed to a determined assault with a bayonet.

"Christ!" said Walker as he opened the first. In awe he pulled out a shimmering sparkle of jewels, a necklace of diamonds and emeralds.

"What's that worth?" Coertze asked Harrison.

Harrison shook his head dumbly. "Gee, I wouldn't know." He smiled faintly. "Not my kind of stone."

They were ransacking the boxes when Coertze pulled out a gold cigarette case. "This one's got an inscription," he said. and read it aloud. "'*Caro Benito da parte di Adolfe—Brennero —1940.*'"

Harrison said slowly, "Hitler had a meeting with Mussolini at the Brenner Pass in 1940. That's when Musso decided to kick in on the German side."

"So now we know who this belongs to," said Walker, waving his hand.

"Or used to belong to," repeated Coertze slowly. "But who does it belong to now?"

They looked at each other.

Coertze broke the silence. "Come on, let's see what's in the last truck."

The fourth truck was full of packing cases containing more papers. But there was one box holding a crown.

Harrison struggled to lift it. "Who's the giant who wears this around the palace?" he asked nobody in particular. The crown was thickly encrusted with jewels—rubies and emeralds, but no diamonds. It was ornate and very heavy. "No wonder they say 'uneasy lies the head that wears a crown,'" cracked Harrison.

He lowered the crown into the box. "Well, what do we do now?"

Coertze scratched his head. "It's quite a problem," he admitted.

"I say we keep it," said Harrison bluntly. "It's ours by right of conquest."

Now it was in the open—the secret thought that no one would admit except the extrovert Harrison. It cleared the air and made things much easier.

Coertze said, "I suppose we must bring in the rest of the boys and vote on it."

"That'll be no good unless it's a unanimous vote," said Harrison almost casually.

They saw his point. If one of them held out in favour of telling the Count, then the majority vote would be useless. At last Walker said, " It may not arise. Let's vote on it and see."

All was quiet on the road so Donato and Parker were brought in from their sentry duty. The prisoners were herded into a truck so that Alberto could join in the discussion, and they settled down as a committee of ways and means.

Harrison needn't have worried—it *was* a unanimous vote. There was too much temptation for it to be otherwise.

" One thing's for sure," said Harrison. " When this stuff disappears there's going to be the biggest investigation ever, no matter who wins the war. The Italian Government will never rest until it's found, especially those papers. I'll bet they're dynamite."

Coertze was thoughtful. " That means we must hide the treasure *and* the trucks. *Nothing* must be found. It must be as though the whole lot has vanished into thin air."

" What are we going to do with it?" asked Parker. He looked at the stony ground and the thin soil. " We might just bury the treasure if we took a week doing it, but we can't even begin to bury one truck, let alone four."

Harrison snapped his fingers. " The old lead mines," he said. " They're not far from here."

Coertze's face lightened. " *Ja*," he said. " There's one winze that would take the lot."

Parker said, " What lead mines—and what's a winze, for God's sake?"

" It's a horizontal shaft driven into a mountain," said Harrison. " These mines have been abandoned since the turn of the century. No one goes near them any more."

Alberto said, " We drive all the trucks inside . . ."

". . . and blow in the entrance," finished Coertze with gusto.

" Why not keep some of the jewels?" suggested Walker.

" No," said Coertze sharply. " It's too dangerous—Harrison is right. There'll be all hell breaking loose when this stuff vanishes for good. Everything must be buried until it's safe to recover it."

" Know any good jewel fences?" asked Harrison sardonically. " Because if you don't, how would you get rid of the jewels?"

They decided to bury everything—the trucks, the bodies, the gold, the papers, the jewels—everything. They restowed the trucks, putting all the valuables into two trucks and all the non-valuables such as the documents into the other two. It was intended to drive the staff car into the tunnel first with the motor-cycle carried in the back, then the trucks carrying papers and bodies, and lastly the trucks with the gold and jewels.

"That way we can get out the stuff we want quite easily," said Coertze.

The disposal of the trucks was easy enough. There was an unused track leading to the mines which diverged off the dusty road they were on. They drove up to the mine and reversed the trucks into the biggest tunnel in the right order. Coertze and Harrison prepared a charge to blow down the entrance, a simple job taking only a few minutes, then Coertze lit the fuse and ran back.

When the dust died down they saw that the tunnel mouth was entirely blocked—making a rich mausoleum for seventeen men.

"What do we tell the Count?" asked Parker.

"We tell him we ran into a little trouble on the way," said Coertze. "Well, we did, didn't we?" He grinned and told them to move on.

When they got back they heard that Umberto had run into trouble and had lost a lot of men. The Communists hadn't turned up and he hadn't had enough machine-guns.

I said, "You mean the gold's still there."

"That's right," said Walker, and hammered his fist on the counter. "Let's have another drink."

I didn't get much out of him after that. His brain was pickled in brandy and he kept wandering into irrelevancies, But he did answer one question coherently.

I asked, "What happened to the two German prisoners?"

"Oh, them," he said carelessly. "They were shot while escaping. Coertze did it."

IV

Walker was too far gone to walk home that night, so I got his address from a club steward, poured him into a taxi and forgot about him. I didn't think much of his story—it was just the maunderings of a drunk. Maybe he had found something in Italy, but I doubted if it was anything big—my imagination boggled at the idea of four truck loads of gold and jewels.

I wasn't allowed to forget him for long because I saw him the following Sunday in the club bar gazing moodily into a brandy glass. He looked up, caught my eye and looked away hastily as though shamed. I didn't go over to speak to him; he wasn't altogether my type—I don't go for drunks much.

Later that afternoon I had just come out of the swimming pool and was enjoying a cigarette when I became aware that Walker was standing beside me. As I looked up, he said awkwardly, "I think I owe you some money—for the taxi fare the other night."

"Forget it," I said shortly.

He dropped on one knee. "I'm sorry about that. Did I cause any trouble?"

I smiled. "Can't you remember?"

"Not a damn' thing," he confessed. "I didn't get into a fight or anything, did I?"

"No, we just talked."

His eyes flickered. "What about?"

"Your experiences in Italy. You told me a rather odd story."

"I told you about the gold?"

I nodded. "That's right."

"I was drunk," he said. "As shickered as a coot. I shouldn't have told you about that. You haven't mentioned it to anyone, have you?"

"No, I haven't," I said. "You don't mean it's true?" He certainly wasn't drunk now.

"True enough," he said heavily. "The stuff's still up there—in a hole in the ground in Italy. I'd not like you to talk about it."

"I won't," I promised.

"Come and have a drink," he suggested.

"No, thanks," I said. "I'm going home now."

He seemed depressed. "All right," he said, and I watched him walk lethargically up to the club house.

After that, he couldn't seem to keep away from me. It was as though he had delivered a part of himself into my keeping and he had to watch me to see that I kept it safe. He acted as though we were both partners in a conspiracy, with many a nod and wink and a sudden change of subject if he thought we were being overheard.

He wasn't so bad when you got to know him, if you discounted the incipient alcoholism. He had a certain charm when he wanted to use it and he most surely set out to charm me. I don't suppose it was difficult; I was a stranger in a strange land and he was company of sorts.

He ought to have been an actor for he had the gift of mimicry. When he told me the story of the gold his mobile face altered plastically and his voice changed until I could see the bull-headed Coertze, gentle Donato and the tougher-fibred Alberto. Although Walker had normally a slight trace of a South African accent, he could drop it at will to take on the heavy gutturals of the Afrikaner or the speed and sibilance of the Italian. His Italian was rapid and fluent and he was probably one of those people who can learn a language in a matter of weeks.

I had lost most of my doubts about the truth of his story. It was too damned circumstantial. The bit about the inscription on the cigarette case impressed me a lot; I couldn't see Walker making up a thing like that. Besides, it wasn't the brandy talking all the time; he still stuck to the same story, which didn't change a fraction under many repetitions—drunk and sober.

Once I said, "The only thing I can't figure is that big crown."

"Alberto thought it was the royal crown of Ethiopia," said Walker. "It wouldn't be worn about the palace—they'd only use it for coronations."

That sounded logical. I said, "How do you know that the others haven't dug up the lot? There's still Harrison and Parker—and it would be dead easy for the two Italians; they're on the spot."

Walker shook his head. "No, there's only Coertze and me. The others were all killed." His lips twisted. "It seemed to be unhealthy to stick close to Coertze. I got scared in the end and beat it."

I looked hard at him. "Do you mean to say that Coertze murdered them?"

"Don't put words in my mouth," said Walker sharply. "I didn't say that. All I know is that four men were killed when they were close to Coertze." He ticked them off on his fingers. "Harrison was the first—that happened only three days after we buried the loot."

He tapped a second finger. "Next came Alberto—I saw that happen. It was as neat an accident as anyone could arrange. Then Parker. He was killed in action just like Harrison, and, just like Harrison, the only person who was anywhere near him was Coertze."

He held up three fingers and slowly straightened the fourth. "Last was Donato. He was found near the camp with his head bashed in. They said he'd been rock-climbing, so the verdict was accidental death—but not in my book. That was enough for me, so I quit and went south."

I thought about this for a while, then said, "What did you mean when you said you saw Alberto killed?"

"We'd been on a raid," said Walker. "It went O.K. but the Germans moved fast and got us boxed in. We had to get out by the back door, and the back door was a cliff. Coertze was good on a mountain and he and Alberto went first, Coertze leading. He said he wanted to find the easiest way down, which was all right—he usually did that.

"He went along a ledge and out of sight, then he came back and gave Alberto the O.K. sign. Then he came back to tell us it was all right to start down, so Parker and I went next. We followed Alberto and when we got round the corner we saw that he was stuck.

"There were no hand holds ahead of him and he'd got himself into a position where he couldn't get back, either. Just as we got there he lost his nerve—we could see him quivering and shaking. There he was, like a fly on the side of that cliff with a hell of a long drop under him and a pack of Germans ready to drop on top of him, and he was shaking like a jelly.

"Parker shouted to Coertze and he came down. There was just room enough for him to pass us, so he said he'd go to help Alberto. He got as far as Alberto and Alberto fell off. I swear that Coertze pushed him."

"Did you see Coertze push him?" I asked.

"No," Walker admitted. "I couldn't see Alberto at all once Coertze had passed us. Coertze's a big bloke and he isn't made of glass. But why did he give Alberto the O.K. sign to go along that ledge?"

"It could have been an honest mistake."

Walker nodded. "That's what I thought at the time. Coertze said afterwards that he didn't mean that Alberto should go as far as that. There *was* an easier way down just short of where Alberto got stuck. Coertze took us down there."

He lit a cigarette. "But when Parker was shot up the following week I started to think again."

"How did it happen?"

Walker shrugged. "The usual thing—you know how it is in a fight. When it was all over we found Parker had a hole in his head. Nobody saw it happen, but Coertze was nearest." He paused. "The hole was in the *back* of the head."

"A German bullet?"

Walker snorted. "Brother, we didn't have time for an autopsy; but it wouldn't have made any difference. We were using German weapons and ammo—captured stuff; and Coertze *always* used German guns; he said they were better than the British." He brooded. "That started me thinking seriously. It was all too pat—all these blokes being knocked off so suddenly. When Donato got his, I quit. The Foreign Legion was just about busted anyway. I waited until the Count had sent Coertze off somewhere, then I collected my gear, said good-bye and headed south to the Allied lines. I was lucky—I got through."

"What about Coertze?"

"He stayed with the Count until the Yanks came up. I saw him in Jo'burg a couple of years ago. I was crossing the road to go into a pub when I saw Coertze going through the door. I changed my mind; I had a drink, but not in *that* pub."

He shivered suddenly. "I want to stay as far from Coertze as I can. There's a thousand miles between Cape Town and Johannesburg—that ought to be enough." He stood up suddenly. "Let's go and have a drink, for God's sake."

So we went and had a drink—several drinks.

V

During the next few weeks I could see that Walker was on the verge of making me a proposition. He said he had some money due to him and that he would need a good friend. At last he came out with it.

"Look," he said. "My old man died last year and I've got two thousand pounds coming when I can get it out of the lawyer's hands. I could go to Italy on two thousand pounds."

"So you could," I said.

He bit his lip. "Hal, I want you to come with me."

"For the gold?"

"That's right; for the gold. Share and share alike."

"What about Coertze?"

"To hell with Coertze," said Walker violently. "I don't want to have anything to do with him."

I thought about it. I was young and full of vinegar in those days, and this sounded just the ticket—if Walker was telling the truth. And if he wasn't telling the truth, why would he finance me to a trip to Italy? It seemed a pleasantly adventurous thing to do, but I hesitated. "Why me?" I asked.

"I can't do it myself," he said. "I wouldn't trust Coertze,

and you're the only other chap who knows anything about it. And I trust you, Hal, I really do."

I made up my mind. "All right, it's a deal. But there are conditions."

"Trot them out."

"This drinking of yours has to stop," I said. "You're all right when you're sober, but when you've got a load on you're bloody awful. Besides, you know you spill things when you're cut."

He rearranged his eager face into a firm expression. "I'll do it, Hal; I won't touch a drop," he promised.

"All right," I said. "When do we start?"

I can see now that we were a couple of naïve young fools. We expected to be able to lift several tons of gold from a hole in the ground without too much trouble. We had no conception of the brains and organisation that would be needed—and were needed in the end.

Walker said, "The lawyer tells me that the estate will be settled finally in about six weeks. We can leave any time after that."

We discussed the trip often. Walker was not too much concerned with the practical difficulties of getting the gold, nor with what we were going to do with it once we had it. He was mesmerised by the millions involved.

He said once, "Coertze estimated that there were four tons of gold. At the present price that's well over a million pounds. Then there's the lire—packing cases full of the stuff. You can get a hell of a lot of lire into a big packing case."

"You can forget the paper money," I said. "Just pass one of those notes and you'll have the Italian police jumping all over you."

"We can pass them outside Italy," he said sulkily.

"Then you'll have to cope with Interpol."

"All right," he said impatiently. "We'll forget the lire. But there's still the jewellery—rings and necklaces, diamonds and emeralds." His eyes glowed. "I'll bet the jewels are worth more than the gold."

"But not as easily disposed of," I said.

I was getting more and more worried about the sheer physical factors involved. To make it worse, Walker wouldn't tell me the position of the lead mine, so I couldn't do any active planning at all.

He was behaving like a child at the approach of Christmas, eager to open his Christmas stocking. I couldn't get him to face facts and I seriously contemplated pulling out of this mad scheme. I could see nothing ahead but a botched job with a probably lengthy spell in an Italian jail.

The night before he was to go to the lawyer's office to sign the final papers and receive his inheritance I went to see him at his hotel. He was half-drunk, lying on his bed with a bottle conveniently near.

"You promised you wouldn't drink," I said coldly.

"Aw, Hal, this isn't drinking; not what I'm doing. It's just a little taste to celebrate."

I said, "You'd better cut your celebration until you've read the paper."

"What paper?"

"This one," I said, and took it from my pocket. "That little bit at the bottom of the page."

He took the paper and looked at it stupidly. "What must I read?"

"That paragraph headed: 'Italians Sentenced.'"

It was only a small item, a filler for the bottom of a page.

Walker was suddenly sober. "But they *were* innocent," he whispered.

"That didn't prevent them from getting it in the neck," I said brutally.

"God!" he said. "They're still looking for it."

"Of course they are," I said impatiently. "They'll keep looking until they find it." I wondered if the Italians were more concerned about the gold or the documents.

I could see that Walker had been shocked out of his euphoric dreams of sudden wealth. He now had to face the fact that pulling gold out of an Italian hole had its dangers.

"This makes a difference," he said slowly. "We can't go now. We'll have to wait until this dies down."

"Will it die down—ever?" I asked.

He looked up at me. "I'm not going now," he said with the firmness of fear. "The thing's off—it's off for a long time."

In a way I was relieved. There was a weakness in Walker that was disturbing and which had been troubling me. I had been uneasy for a long time and had been very uncertain of the wisdom of going to Italy with him. Now it was decided.

I left him abruptly in the middle of a typical action—pouring another drink.

As I walked home one thought occurred to me. The newspaper report confirmed Walker's story pretty thoroughly. That was something.

VI

It was long past lunch-time when I finished the story. My throat was dry with talking and Jean's eyes had grown big and round.

"It's like something from the Spanish Main," she said. "Or a Hammond Innes thriller. Is the gold still there?"

I shrugged. "I don't know. I haven't read anything about it in the papers. For all I know it's still there—if Walker or Coertze haven't recovered it."

"What happened to Walker?"

"He got his two thousand quid," I said. "Then embarked on a career of trying to drink the distilleries dry. It wasn't long before he lost his job and then he dropped from sight. Someone told me he'd gone to Durban. Anyway, I haven't seen him since."

Jean was fascinated by the story and after that we made a game of it, figuring ways and means of removing four tons of gold from Italy as unobtrusively as possible. Just as an academic exercise, of course. Jean had a fertile imagination and some of her ideas were very good.

Because that was the problem—the sheer physical difficulty of moving four tons of gold without anyone being aware that it was even there to be moved.

In 1959 we got clear of our indebtedness to the bank by dint of strict economy. The yard was ours now with no strings attached and we celebrated by laying the keel of a 15-tonner I had designed for Jean and myself. My old faithful *King Penguin,* one of the first of her class, was all right for coastal pottering, but we had the idea that one day we would do some ocean voyaging, and we wanted a bigger boat.

A 15-tonner is just the right size for two people to handle and big enough to live in indefinitely. This boat was to be forty feet overall, thirty feet on the waterline with eleven feet beam. She would be moderately canvased for ocean voyaging and would have a big auxiliary diesel engine. We were going to call her *Sanford* in memory of old Tom.

When she was built we would take a year's leave, sail north to spend some time in the Mediterranean, and come back by the east coast, thus making a complete circumnavigation of Africa. Jean had a mischievous glint in her eye. "Perhaps we'll bring that gold back with us," she said.

But two months later the blow fell.

I had designed a boat for Bill Meadows and had sent him the drawings for approval. By mishap the accommodation plans had been left out of the packet, so Jean volunteered to take them to Fish Hoek where Bill lives.

It's a nice drive to Fish Hoek along the Chapman's Peak road with views of sea and mountain, far better than anything I have since seen on the Riviera. Jean delivered the drawings and on the way back in the twilight a drunken oaf in a high-powered American car forced her off the road and she fell three hundred feet into the sea.

The bottom dropped out of my life.

It meant nothing to me that the driver of the other car got five years for manslaughter—that wouldn't bring Jean back. I let things slide at the yard and if it hadn't been for Harry Marshall the business would have gone to pot.

It was then that I tallied up my life and made a sort of mental balance sheet. I was thirty-six years old; I had a good business which I had liked but which now I didn't seem to like so much; I had my health and strength—boat-building and sailing tend to keep one physically fit—and I had no debts. I even had money in the bank with more rolling in all the time.

On the other side of the balance sheet was the dreadful absence of Jean, which more than counter-balanced all the advantages.

I felt I couldn't stay at the yard or even in Cape Town, where memories of Jean would haunt me at every corner. I wanted to get away. I was waiting for something to happen. I was ripe for mischief.

VII

A couple of weeks later I was in a bar on Adderley Street having a drink or three. It wasn't that I'd taken to drink, but I was certainly drinking more than I had been accustomed to. I had just started on my third brandy when I felt a touch at my elbow and a voice said, "Hallo, I haven't seen you for a long time."

I turned and found Walker standing next to me.

The years hadn't dealt kindly with Walker. He was thinner, his dark, good looks had gone to be replaced by a sharpness of feature, and his hairline had receded. His clothes were unpressed and frayed at the edges, and there was an air of seediness about him which was depressing.

"Hallo," I said. "Where did you spring from?" He was

looking at my full glass of brandy, so I said, "Have a drink."

"Thanks,' he said quickly. "I'll have a double."

That gave me a pretty firm clue as to what had happened to Walker, but I didn't mind being battened upon for a couple of drinks, so I paid for the double brandy.

He raised the glass to his lips with a hand that trembled slightly, took a long lingering gulp, then put the glass down, having knocked back three-quarters of the contents. "You're looking prosperous," he said.

"I'm not doing too badly."

He said, "I was sorry to hear of what happened to your wife." He hurried on as he saw my look of inquiry. "I read about it in the paper. I thought it must have been your wife—the name was the same and all that."

I thought he had spent some time hunting me up. Old friends and acquaintances are precious to an alcoholic; they can be touched for the odd drink and the odd fiver.

"That's finished and best forgotten," I said shortly. Unwittingly, perhaps, he had touched me on the raw—he had brought Jean back. "What are you doing now?"

He shrugged. "This and that."

"You haven't picked up any gold lately?" I said cruelly. I wanted to pay him back for putting Jean in my mind.

"Do I look as though I have?" he asked bitterly. Unexpectedly, he said, "I saw Coertze last week."

"Here—in Cape Town?"

"Yes. He'd just come back from Italy. He's back in Jo'burg now, I expect."

I smiled. "Did *he* have any gold with him?"

Walker shook his head. "He said that nothing's changed." He suddenly gripped me by the arm. "The gold's still there—nobody's found it. It's still there—four tons of gold in that tunnel—and all the jewels." He had a frantic urgency about him.

"Well, why doesn't he do something about it?" I said. "Why doesn't he go and get it out? Why don't you both go?"

"He doesn't like me," said Walker sulkily. "He'll hardly speak to me. He took one of my cigarettes from the packet on the counter, and I lit it for him, amusedly. "It isn't easy to get it out of the country," he said. "Even Sergeant High-and-Mighty Coertze hasn't found a way."

He grinned tightly. "Imagine that," he said, almost gaily. "Even the brainy Coertze can't do it. He put the gold in a hole in the ground and he's too scared to get it out." He began to laugh hysterically.

I took his arm. "Take it easy."

His laughter choked off suddenly. "All right," he said. "Buy me another drink; I left my wallet at home."

I crooked my finger at the bartender and Walker ordered another double. I was beginning to understand the reason for his degradation. For fourteen years the knowledge that a fortune in gold was lying in Italy waiting to be picked up had been eating at him like a cancer. Even when I knew him ten years earlier I was aware of the fatal weakness in him, and now one could see that the bitterness of defeat had been too much. I wondered how Coertze was standing up to the strain. At least he seemed to be doing something about it, even if only keeping an eye on the situation.

I said carefully, "If Coertze was willing to take you, would you be prepared to go to Italy to get the stuff out?"

He was suddenly very still. "What d'you mean?" he demanded. "Have you been talking to Coertze?"

"I've never laid eyes on the man."

Walker's glance shifted nervously about the bar, then he straightened. "Well, if he . . . wanted me; if he . . . needed me—I'd be prepared to go along." He said this with bravado but the malice showed through when he said, "He needed me once, you know; he needed me when we buried the stuff."

"You wouldn't be afraid of him?"

"What do you mean—afraid of him? Why should I be afraid of him? I'm afraid of nobody."

"You were pretty certain he'd committed at least four murders."

He seemed put out. "Oh, that! That was a long time ago. And I never said he'd murdered anybody. I never said it."

"No, you never actually said it."

He shifted nervously on the bar stool. "Oh, what's the use? He won't ask me to go with him. He said as much last week."

"Oh, yes, he will," I said softly.

Walker looked up quickly. "Why should he?"

I said quietly, "Because I know a way of getting that gold out of Italy and of taking it anywhere in the world, quite simply and relatively safely."

His eyes widened. "What is it? How can you do it?"

"I'm not going to tell you," I said equably. "After all, you wouldn't tell me where the gold's hidden."

"Well, let's do it," he said. "I'll tell you where it is, you get it out, and Bob's your uncle. Why bring Coertze into it?"

"It's a job for more than two men," I said. "Besides, he deserves a share—he's been keeping an eye on the gold for fourteen years, which is a damn' sight more than you've been doing." I failed to mention that I considered Walker the weakest of reeds. "Now, how will you get on with Coertze if this thing goes through?"

He turned sulky. "All right, I suppose, if he lays off me. But I won't stand for any of his sarcasm." He looked at me in wonder as though what we were talking about had just sunk in. "You mean there's a chance we can get the stuff out—a real chance?"

I nodded and got off the bar stool. "Now, if you'll excuse me."

"Where are you going?" he asked quickly.

"To phone the airline office," I said. "I want a seat on to-morrow's Jo'burg plane. I'm going to see Coertze."

The sign I had been waiting for had arrived.

II. COERTZE

Air travel is wonderful. At noon the next day I was booking into a hotel in Johannesburg, a thousand miles from Cape Town.

On the plane I had thought a lot about Coertze. I had made up my mind that if he didn't bite then the whole thing was off—I couldn't see myself relying on Walker. And I had to decide how to handle him—from Walker's account he was a pretty tough character. I didn't mind that; I could be tough myself when the occasion arose, but I didn't want to antagonise him. He would probably be as suspicious as hell, and I'd need kid gloves.

Then there was another thing—the financing of the expedition. I wanted to hang on to the boatyard as insurance in case this whole affair flopped, but I thought if I cut Harry Marshall in for a partnership in the yard, sold my house and my car and one or two other things, I might be able to raise about £25,000—not too much for what I had in mind.

But it all depended on Coertze. I smiled when I considered where he was working. He had a job in the Central Smelting Plant which refined gold from all the mines on the Reef. More gold had probably passed through his hands in the last few years than all the Axis war-lords put together had buried throughout the world.

It must have been tantalising for him.

I phoned the smelting plant in the afternoon. There was a pause before he came on the line. "Coertze," he said briefly.

I came to the point. "My name's Halloran," I said, "A mutual friend—Mr. Walker of Cape Town—tells me you have been experiencing difficulty in arranging for the delivery of goods from Italy. I'm in the import-export business; I thought I might be able to help you."

A deep silence bored into my ear.

I said, "My firm is fully equipped to do this sort of work. We never have much trouble with the Customs in cases like this."

It was like dropping a stone into a very deep well and listening for the splash.

"Why don't you come to see me," I said. "I don't want to take up your time now; I'm sure you're a busy man. Come at seven this evening and we'll discuss your difficulties over dinner. I'm staying at the Regency—it's in Berea, in . . ."

"I know where it is," said Coertze. His voice was deep and harsh with a guttural Afrikaans accent.

"Good; I'll be expecting you," I said, and put down the phone.

I was pleased with this first contact. Coertze was suspicious and properly so—he'd have been a fool not to be. But if he came to the hotel he'd be hooked, and all I had to do would be to jerk on the line and set the hook in firmly.

I was pretty certain he'd come; human curiosity would see to that. If he didn't come, then he wouldn't be human—or he'd be superhuman.

He came, but not at seven o'clock and I was beginning to doubt my judgment of the frailty of human nature. It was after eight when he knocked on the door, identified me, and said, "We'll forget the dinner; I've eaten."

"All right," I said. "But what about a drink?" I crossed the room and put my hand on the brandy bottle. I was pretty certain it would be brandy—most South Africans drink it.

"I'll have a Scotch,' 'he said unexpectedly. "Thanks," he added as an afterthought.

As I poured the drinks I glanced at him. He was a bulky man, broad of chest and heavy in the body. His hair was black and rather coarse and he had a shaggy look about him. I'd bet that when stripped he'd look like a grizzly bear. His eyebrows were black and straight over eyes of a snapping electric blue. He had looked after himself better than Walker; his belly was flat and there was a sheen of health about him.

I handed him a drink and we sat down facing each other. He was tense and wary, although he tried to disguise it by over-relaxing in his chair. We were like a couple of duellists who have just engaged blades.

"I'll come to the point," I said. "A long time ago Walker told me a very interesting story about some gold. That was ten years ago and we were going to do something about it, but it didn't pan out. That might have been lucky because we'd have certainly made a botch of the job."

I pointed my finger at him. "You've been keeping an eye on it. You've probably popped across to Italy from time to time just to keep your eye on things in general. You've been racking your brains trying to think of a way of getting that gold out of Italy, but you haven't been able to do it. You're stymied."

His face had not changed expression; he would have made a good poker player. He said, "When did you see Walker?"

"Yesterday—in Cape Town."

The craggy face broke into a derisive grin. "And you flew up to Jo'burg to see me just because a *dronkie* like Walker told you a cock-and-bull story like that? Walker's a no-good hobo; I see a dozen like him in the Library Gardens every day," he said contemptuously.

"It's not a cock-and-bull story, and I can prove it."

Coertze just sat and looked at me like a stone gargoyle, the whisky glass almost lost in his huge fist.

I said, "What are you doing here—in this room? If there was no story, all you had to do was to ask me what the hell I was talking about when I spoke to you on the phone. The fact that you're here proves there's something in it."

He made a fast decision. "All right," he said. "What's your proposition?"

I said, "You still haven't figured a way of moving four tons of gold out of Italy. Is that right?"

He smiled slowly. "Let's assume so," he said ironically.

"I've got a foolproof way."

He put down his glass and produced a packet of cigarettes. "What is it?"

"I'm not going to tell you—yet."

He grinned. "Walker hasn't told you where the gold is, has he?"

"No, he hasn't," I admitted. "But he would if I put pressure on him. Walker can't stand pressure; you know that."

"He drinks too much," said Coertze. "And when he drinks he talks; I'll bet that's how he came to spill his guts to you." He lit his cigarette. "What do you want out of it?"

"Equal shares," I said firmly. "A three-way split after all expenses have been paid."

"And Walker comes with us on the job. Is that right?"

"Yes," I said.

Coertze moved in his chair. "Man, it's like this," he said. "I don't know if you've got a foolproof way of getting the gold out or if you haven't. I thought *I* had it licked a couple of times. But let's assume your way is going to work. Why should we take Walker?"

He held up his hand. "I'm not suggesting we do him down or anything like that—although he'd think nothing of cheating us. Give him his share after it's all over, but for God's sake keep him out of Italy. He'll make a balls-up for sure."

I thought of Harrison and Parker and the two Italians. "You don't seeem to like him."

Coertze absently fingered a scar on his forehead. "He's unreliable," he said. "He almost got me killed a couple of times during the war."

I said, "No, we take Walker. I don't know for certain if three of us can pull it off, and with two it would be impossible. Unless you want to let someone else in?"

He smiled humourlessly. "That's not on—not with you coming in. But Walker had better keep his big mouth shut from now on."

"Perhaps it would be better if he stopped drinking," I suggested.

"That's right," Coertze agreed. "Keep him off the pots. A few beers are all right, but keep him off the hard-tack. That'll be your job; I don't want to have anything to do with the rat."

He blew smoke into the air, and said, "Now let's hear your proposition. If it's good, I'll come in with you. If I don't think it'll work, I won't touch it. In that case, you and Walker can do what you damn' well like, but if you go for that gold you'll have me to reckon with. I'm a bad bastard when I'm crossed."

"So am I," I said.

We grinned at each other. I liked this man, in a way. I wouldn't trust him any more than I'd trust Walker, but I had the feeling that while Walker would stick a knife in your back, Coertze would at least shoot you down from the front.

"All right," he said. "Let's have it."

"I'm not going to tell you—not here in this room." I saw his expression and hurried on. "It isn't that I don't trust you, it's simply that you wouldn't believe it. You have to see it—and you have to see it in Cape Town."

He looked at me for a long moment, then said, "All right, if that's the way you want it, I'll play along." He paused to think. "I've got a good job here, and I'm not going to give it up on your say-so. There's a long week-end coming up—that gives me three days off. I'll fly down to Cape Town to see what you have to show me. If it's good, the job can go hang; if it isn't, then I've still got the job."

"I'll pay your fare," I said.

"I can afford it," he grunted.

"If it doesn't pan out, I'll pay your fare," I insisted. "I wouldn't want you to be out of pocket."

He looked up and grinned. "We'll get along," he said. "Where's that bottle?"

As I was pouring another couple of drinks, he said, "You said you were going to Italy with Walker. What stopped you?"

I took the clipping from my pocket and passed it to him. He read it and laughed. "That must have scared Walker. I was there at the time," he said unexpectedly.

"In Italy?"

He sipped the Scotch and nodded. "Yes; I saved my army back-pay and my gratuity and went back in '48. As soon as I got there all hell started popping about this trial. I read about it in the papers and you never heard such a lot of bull in your life. Still, I thought I'd better lie low, so I had a *lekker* holiday with the Count."

"With the Count?" I said in surprise.

"Sure," he said. "I stay with the Count every time I go to Italy. I've been there four times now."

I said, "How did you reckon to dispose of the gold once you got it out of Italy?"

"I've got all that planned," he said confidently. "They're always wanting gold in India and you get a good price. You'd be surprised at the amount of gold smuggled out of this country in small packets that ends up in India."

He was right—India is the gold sink of the world—but I said casually, "My idea is to go the other way—to Tangier. It's an open port with an open gold market. You should be able to sell four tons of gold there quite easily—and it's legal, too. No trouble with the police."

He looked at me with respect. "I hadn't thought of that. I don't know much about this international finance."

"There's a snag," I said. "Tangier is closing up shop next year; it's being taken over by Morocco. Then it won't be a free port any more and the gold market will close."

"When next year?"

"April 19," I said. "Nine months from now. I think we'll just about have enough time."

He smiled. "I never thought about selling the gold legally; I didn't think you could. I though the governments had got all that tied up. Maybe I should have met you sooner."

"It wouldn't have done you any good," I said. "I hadn't the brains then that I have now."

He laughed and we proceeded to kill the bottle.

II

Coertze came down to Cape Town two weeks later. I met him at the airport and drove him directly to the yard, where Walker was waiting.

Walker seemed to shrink into himself when I told him that Coertze was visiting us. In spite of his braggart boasts, I could see he didn't relish close contact. If half of what he had said about Coertze was true, then he had every reason to be afraid.

Come to think of it—so had I!

It must have been the first time that Coertze had been in a boatyard and he looked about him with keen interest and asked a lot of questions, nearly all of them sensible. At last, he said, "Well, what about it?"

I took them down to the middle slip where Jimmy Murphy's *Estralita* was waiting to be drawn up for an overhaul. "That's a sailing yacht," I said. "A 15-tonner. What would you say her draft is—I mean, how deep is she in the water?"

Coertze looked her over and then looked up at the tall mast. "She'll need to be deep to counter-balance that lot," he said. "But I don't know how much. I don't know anything about boats."

Considering he didn't know anything about boats, it was a very sensible answer.

Her draft is six feet in normal trim," I said. "She's drawing less now because a lot of gear has been taken out of her."

His eyes narrowed. "I'd have thought it would be more

than that," he said. "What happens when the wind blows hard on the sails? Won't she tip over?"

This was going well and Coertze was on the ball. I said, "I have a boat like this just being built, another 15-tonner. Come and have a look at her."

I led the way up to the shed where *Sanford* was being built and Coertze followed, apparently content that I was leading up to a point. Walker tagged on behind.

I had pressed to get *Sanford* completed and she was ready for launching as soon as the glass-fibre sheathing was applied and the interior finished.

Coertze looked up at her. "They look bloody big out of the water," he commented.

I smiled. That was the usual lay reaction. "Come aboard," I said.

He was impressed by the spaciousness he found below and commented favourably on the way things were arranged. "Did you design all this?" he asked.

I nodded.

"You could live in here, all right." he said, inspecting the galley.

"You could—and you will," I said. "This is the boat in which we're going to take four tons of gold out of Italy."

He looked surprised and then he frowned. "Where are you going to put it?"

I said, "Sit down and I'll tell you something about sailing boats you don't know." Coertze sat uncomfortably on the edge of the starboard settee which had no mattress as yet, and waited for me to explain myself.

"This boat displaces—weighs, that is—ten tons, and . . ."

Walker broke in. "I thought you said she was a 15-tonner."

"That's Thames measure—yacht measure. Her displacement is different."

Coertze looked at Walker. "Shut up and let the man speak." He turned to me. "If the boat weighs ten tons and you add another four tons, she'll be pretty near sinking, won't she? And where are you going to put it? It can't be out in the open where the cops can see it."

I said patiently, "I said I'd tell you something about sailing boats that you didn't know. Now, listen—about forty per cent of the weight of any sailing boat is ballast to keep her the right way up when the wind starts to press on those sails."

I tapped the cabin sole with my foot. "Hanging on the bottom of this boat is a bloody great piece of lead weighing precisely four tons."

Coertze looked at me incredulously, a dawning surmise in his eyes. I said, "Come on, I'll show you."

We went outside and I showed them the lead ballast keel. I said, "All this will be covered up next week because the boat will be sheathed to keep out the marine borers."

Coertze was squatting on his heels looking at the keel. "This is it," he said slowly. "This is it. The gold will be hidden under water—built in as part of a boat." He began to laugh, and after a while Walker joined in. I began to laugh, too, and the walls of the shed resounded.

Coertze sobered suddenly. "What's the melting point of lead?" he asked abruptly.

I knew what was coming. "Four-fifty degrees centigrade," I said. "We've got a little foundry at the top of the yard where we pour the keels."

"*Ja*," he said heavily. "You can melt lead on a kitchen stove. But gold melts at over a thousand centigrade and we'll need more than a kitchen stove for that. I *know;* melting gold is my job. Up at the smelting plant we've got bloody big furnaces."

I said quickly, "I've thought of that one, too. Come up to the workshop—I'll show you something else you've never seen before."

In the workshop I opened a cupboard and said, "This gadget is brand new—just been invented." I hauled out the contraption and put it on the bench. Coertze looked at it uncomprehendingly.

There wasn't much to see; just a metal box, eighteen inches by fifteen inches by nine inches, on the top of which was an asbestos mat and a Heath Robinson arrangement of clamps.

I said, "You've heard of instant coffee—this is instant heat." I began to get the machine ready for operation. "It needs cooling water at at least five pounds an inch pressure—that we get from an ordinary tap. It works on ordinary electric current, too, so you can set it up anywhere."

I took the heart of the machine from a drawer. Again, it wasn't much to look at; just a piece of black cloth, three inches by four. I said, "Some joker in the States discovered how to spin and weave threads of pure graphite, and someone else discovered this application."

I lifted the handle on top of the machine, inserted the graphite mat, and clamped it tight. Then I took a bit of metal and gave it to Coertze.

He turned it in his fingers and said, "What is it?"

"Just a piece of ordinary mild steel. But if this gadget can melt steel, it can melt gold. Right?"

He nodded and looked at the machine dubiously—it wasn't very impressive.

I took the steel from his fingers and dropped it on to the graphite mat, then I gave Walker and Coertze a pair of welders' goggles each. "Better put these on: it gets a bit bright."

We donned the goggles and I switched on the machine. It was a spectacular display. The graphite mat flashed instantly to a white heat and the piece of steel glowed red, then yellow and finally white. It seemed to slump like a bit of melting wax and in less than fifteen seconds it had melted into a little pool. All this to the accompaniment of a violent shower of sparks as the metal reacted with the air.

I switched off the machine and removed my goggles. "We won't have all these fireworks when we melt gold; it doesn't oxidise as easily as iron."

Coertze was staring at the machine. "How does it do that?"

"Something like a carbon arc," I said. "You can get temperatures up to five thousand degrees centigrade. It's only intended to be a laboratory instrument, but I reckon we can melt two pounds of gold at a time. With three of these gadgets and a hell of a lot of spare mats we should be able to work pretty fast."

He said doubtfully, "If we can only pour a couple of pounds at a time, the keel is going to be so full of cracks and flaws that I'm not sure it won't break under its own weight."

"I've thought of that one, too," I said calmly. "Have you ever watched anyone pour reinforced concrete?"

He frowned and then caught on, snapping his fingers.

"We make the mould and put a mesh of wires inside," I said. "That'll hold it together."

I showed him a model I had made, using fuse wire and candle wax, which he examined carefully. "You've done a hell of a lot of thinking about this," he said at last.

"Somebody has to," I said. "Or that gold will stay where it is for another fourteen years."

He didn't like that because it made him appear stupid; but there wasn't anything he could do about it. He started to say something and bit it short, his face flushing red. Then he took a deep breath and said, "All right, you've convinced me. I'm in."

Then *I* took a deep breath—of relief.

III

That night we had a conference.

I said, "This is the drill. *Sanford*—my yacht—will be ready for trials next week. As soon as the trials are over you two are going to learn how to sail under my instruction. In under four months from now we sail for Tangier."

"Christ!" said Walker. "I don't know that I like the sound of that."

"There's nothing to it," I said. "Hundreds of people are buzzing about the Atlantic these days. Hell, people have gone round the world in boats a quarter the size."

I looked at Coertze. "This is going to take a bit of financing. Got any money?"

"About a thousand," he admitted.

"That gets tossed into the kitty," I said. "Along with my twenty-five thousand."

"*Magtig*," he said. "That's a hell of a lot of money."

"We'll need every penny of it," I said. "We might have to buy a small boatyard in Italy if that's the only way we can cast the keel in secrecy. Besides, I'm lending it to the firm of Walker, Coertze and Halloran at one hundred per cent interest. I want fifty thousand back before the three-way split begins. You can do the same with your thousand."

"That sounds fair enough," agreed Coertze.

I said, "Walker hasn't any money and once you've thrown your thousand in the kitty, neither have you. So I'm putting you both on my payroll. You've got to have your smokes and three squares a day while all this is going on."

This bit of information perked Walker up considerably. Coertze merely nodded in confirmation. I looked hard at Walker. "And you stay off the booze or we drop you over the side. Don't forget that."

He nodded sullenly.

Coertze said, "Why are we going to Tangier first?"

"We've got to make arrangements to remelt the gold into standard bars," I said. "I can't imagine any banker calmly taking a golden keel into stock. Anyway, that's for the future; right now I have to turn you into passable seamen—we've got to get to the Mediterranean first."

I took *Sanford* on trials and Walker and Coertze came along for the ride and to see what they were letting themselves in for. She turned out to be everything I've ever wanted in a boat. She was fast for a deep-sea cruiser and not too tender. With a little sail adjustment she had just the right amount of helm and I could see she was going to be all right without any drastic changes.

As we went into a long reach she picked up speed and went along happily with the water burbling along the lee rail and splashing on deck. Walker, his face a little green, said, "I thought you said a keel would hold this thing upright." He was hanging tightly on to the side of the cockpit.

I laughed. I was happier than I had been for a long time. "Don't worry about that. That's not much angle of heel. She won't capsize."

Coertze didn't say anything—he was busy being sick.

The next three months were rough and tough. People forget that the Cape was the Cape of Storms before some early public relations officer changed the name to the Cape of Good Hope. When the Berg Wind blows it can be as uncomfortable at sea as anywhere in the world.

I drove Walker and Coertze unmercifully. In three months I had to turn them into capable seamen, because *Sanford* was a bit too big to sail single-handed. I hoped that the two of them would equal one able-bodied seaman. It wasn't as bad as it sounds because in those three months they put in as much sea time as the average week-end yachtsman gets in three years, and they had the dubious advantage of having a pitiless instructor.

Shore time was spent in learning the theory of sail and the elements of marlin-spike seamanship—how to knot and splice, mend a sail and make baggywrinkle. They grumbled a little at the theory, but I silenced that by asking them what they'd do if I was washed overboard in the middle of the Atlantic.

Then we went out to practise what I had taught—at first in the bay and then in the open sea, cruising coastwise around the peninsula at first, and then for longer distances well out of sight of land.

I had thought that Coertze would prove to be as tough at sea as apparently he was on land. But he was no sailor and never would be. He had a queasy stomach and couldn't stand the motion, so he turned out to be pretty useless at boat handling. But he was hero enough to be our cook on the longer voyages, a thankless job for a sea-sick man.

I would hear him swearing below when the weather was rough and a pot of hot coffee was tossed in his lap. He once told me that he now knew what poker dice felt like when they were shaken in the cup. He wouldn't have stood it for any lesser reason, but the lust for gold was strong in him.

Walker was the real surprise. Coertze and I had weaned him from his liquor over many protests, and he was now eating more and the air and exercise agreed with him. He put on weight, his thin cheeks filled out and his chest broadened. Nothing could replace the hair he had lost, but he seemed a lot more like the handsome young man I had known ten years earlier.

More surprisingly, he turned out to be a natural sailor. He liked *Sanford* and she seemed to like him. He was a good helmsman and could lay her closer to the wind than I could when we were beating to windward. At first I was hesitant to give him a free hand with *Sanford,* but as he proved himself I lost my reluctance.

At last we were ready and there was nothing more to wait for. We provisioned *Sanford* and set sail for the north on November 12, to spend Christmas at sea. Ahead of us was a waste of water with the beckoning lure of four tons of gold at the other side.

I suppose one *could* have called it a pleasure cruise!

III. TANGIER

Two months later we sailed into Tangier harbour, the "Q" flag hoisted, and waited for the doctor to give us pratique and for the Customs to give us the once-over. To port of *Sanford* was the modern city with its sleek, contemporary buildings sharply outlined against the sky. To starboard was the old city—the Arab city—squat and low-roofed and hugging a hill, the skyline only broken by the up-flung spear of a minaret.

To port—Europe; to starboard—Africa.

This was nothing new to Walker and Coertze. They had sown a few wild oats in their army days, roistering in Cairo and Alexandria. On the voyage from Cape Town they had talked much about their army days—and all in Italian, too. We made it a rule to speak as much Italian as possible, and while the others were on a refresher course, I didn't lag far behind even though I had to start from scratch.

We had settled on a good cover story to veil our activities in the Mediterranean. I was a South African boat builder on a cruise combining business with pleasure. I was thinking of expanding into the lucrative Mediterranean market and might buy a boatyard if the price and conditions were right. This story had the advantage of not departing too far from the truth and would serve if we really had to buy a yard to cast the golden keel.

Coertze was a mining man with medical trouble. His doctor had advised him to take a leisurely holiday and so he was crewing *Sanford* for me. His cover story would account for any interest he might take in derelict lead mines.

Walker, who proved to be something of an actor, was a moderately wealthy playboy. He had money but disliked work and was willing to go a long way to avoid it. He had come on this Mediterranean trip because he was bored with South Africa and wanted a change. It was to be his job to set things up in Tangier; to acquire a secluded house where we could complete the last stages of the operation.

All in all, I was quite satisfied, even though I had got a bit tired of Coertze on the way north. He didn't like the way I seemed to be taking charge of things and I had to ram home very forcibly the fact that a ship can only have one skipper. He had seen the point when we ran into heavy weather off the Azores, and it galled him that the despised Walker was the better seaman.

Now we were in Tangier, he had recovered his form and was a bit more inclined to throw his weight around. I could see that I'd have to step on him again before long.

Walker looked about the yacht basin. "Not many sailing boats here," he commented.

That was true. There were a few ungainly-looking fishing boats and a smart ketch, probably bound for the Caribbean. But there were at least twenty big power craft, fast-looking boats, low on the water. I knew what they were.

This was the smuggling fleet. Cigarettes to Spain, cigarette lighters to France, antibiotics to where they could make a profit (although that trade had fallen off), narcotics to everywhere. I wondered if there was much arms smuggling to Algeria.

At last the officials came and went, leaving gouges in my planking from their hob-nailed boots. I escorted them to their launch, and as soon as they had left, Walker touched my arm.

"We've got another visitor," he said.

I turned and saw a boat being sculled across the harbour. Walker said, "He was looking at us through glasses from that boat across there." He pointed to onc of the motor craft. "Then he started to come here."

I watched the approaching dinghy. A European was rowing and I couldn't see his face, but as he dexterously backed-water and swung round to the side of *Sanford* he looked up and I saw that it was Metcalfe.

Metcalfe is one of that international band of scallywags of whom there are about a hundred in the world. They are soldiers of fortune and they flock to the trouble spots, ignoring the danger and going for the money. I was not really surprised to see Metcalfe in Tangier; it had been a pirates' stronghold from time immemorial and would be one of Metcalfe's natural hang-outs.

I had known him briefly in South Africa but I didn't know what he was at the time. All that I knew was that he was a damned good sailor who won a lot of dinghy races at Cape Town and who came close to winning the South African dinghy championship. He bought one of my Falcons and had spent a lot of time at the yard tuning it.

I had liked him and had crewed for him a couple of times. We had had many a drink together in the yacht club bar and he had spent a week-end at Kirstenbosch with Jean and myself. It was in the way of being a firmly ripening friendship between us when he had left South Africa a hop, skip and a jump ahead of the police, who wanted to nail him on a charge of I.D.B. Since then I had not seen him, but I had heard passing mentions and had occasionally seen his name in the papers, usually quoted as being in trouble in some exotic hot-spot.

Now he was climbing on to the deck of *Sanford.*

"I thought it was you," he said. "So I got the glasses to make sure. What are you doing here?"

"Just idly cruising," I said. "Combining business with pleasure. I thought I might see what the prospects in the Med. are like.

He grinned. "Brother, they're good. But that's not in your line, is it?"

I shook my head, and said, "Last I heard of you, you were in Cuba."

"I was in Havana for a bit," he said. "But that was no place for me. It was an *honest* revolution, or at least it was until the Commies moved in. I couldn't compete with them, so I quit."

"What are you doing now?"

He smiled and looked at Walker. "I'll tell you later."

I said, "This is Walker and this is Coertze." There was handshaking all round and Metcalfe said, "It's good to hear a South African accent again. You'd have a good country there if the police weren't so efficient."

He turned to me. "Where's Jean?"

"She's dead," I said. "She was killed in a motor smash."

"How did it happen?"

So I told him of Chapman's Peak and the drunken driver and the three-hundred-foot fall to the sea. As I spoke his face hardened, and when I had finished, he said, "So the bastard only got five years, and if he's a good boy he'll be out in three and a half."

He rubbed his finger against the side of his nose. "I liked Jean," he said. "What's the bludger's name? I've got friends in South Africa who can see to him when he comes out."

"Forget it," I said. "That won't bring Jean back."

He nodded, then slapped his hands together. "Now you're all staying with me at my place; I've got room enough for an army."

I said hesitantly, "What about the boat?"

He smiled. "I see you've heard stories about the Tangier dock thieves. Well, let me tell you they're all true. But that doesn't matter; I'll put one of my men on board. Nobody steals from my men—or me."

He rowed back across the harbour and presently returned with a scar-faced Moroccan, to whom he spoke in quick and guttural Arabic. Then he said, "That's all fixed. I'll have the word passed round the docks that you're friends of mine. Your boat's safe enough, as safe as though it lay in your own yard."

I believed him. I could believe he had a lot of pull in a place like Tangier.

"Let's go ashore," he said. "I'm hungry."

"So am I," said Coertze.

"It'll be a relief not to do any more cooking for a while, won't it?" I said.

"Man," said Coertze. "I wouldn't mind if I never saw a fry-pan again."

"That's a pity," said Metcalfe. "I was looking forward to you making me some *koeksusters*; I always liked South African grub." He roared with laughter and slapped Coertze on the back.

Metcalfe had a big apartment on the Avenida de España, and he gave me a room to myself while Coertze and Walker shared a room. He stayed and chatted while I unpacked my bag.

"South Africa too quiet for you?" he asked.

I went into my carefully prepared standard talk on the reasons I had left. I had no reason to trust Metcalfe more than anyone else—probably less—judging by the kind of man he was. I don't know whether he believed me or not, but he agreed that there was scope in the Mediterranean for a good boatyard.

"You may not get as many commissions to build," he said. "But there certainly is room for a good servicing and maintenance yard. I'd go east, towards Greece, if I were you. The yards in the islands cater mostly for the local fishermen; there's room for someone who understands yachts and yachtsmen."

"What have you got a boat for?" I asked banteringly. "Hiring it out for charter cruises?"

He grinned. "Aw, you know me. I carry all sorts of cargoes; anything except narcotics." He pulled a face. "I'm a bad bastard, I know, but I draw the line at drugs. Anything else I'm game for."

"Including guns to Algeria," I hazarded.

He laughed. "The French in Algiers hate my guts—they

tried to do me down a couple of months ago. I'd unloaded a cargo into some fishing boats and then I ran into Algiers to refuel. I was clean, see! they couldn't touch me—my papers were in order and everything.

"I let the crew go ashore for a drink and I turned in and had a zizz. Then something woke me up—I heard a thump and then a queer noise that seemed to come from *underneath* the boat. So I got up and had a look around. When I got on deck I saw a boat pulling away and there seemed to be a man in the water, swimming alongside it."

He grinned. "Well, I'm a careful and cautious man, so I got my snorkel and my swim-fins and went over the side to have a look-see. What do you think those French Security bastards had done to me?"

I shook my head. "I wouldn't know."

"They'd put a limpet mine on my stern gear. They must have reckoned that if they couldn't nail me down legally they'd do it illegally. If that thing went off it would blow the bottom out of my stern. Well, I got it off the boat and did a bit of heavy thinking. I knew they wouldn't have timed it to blow up in harbour—it wouldn't have looked nice—so I reckoned it was set to blow after I left.

"I slung it round my neck by the cord and swam across the harbour to where the police patrol boat was lying and stuck it under their stern. Let them have the trouble of buying a new boat.

"Next day we left early as planned and, as we moved out, I heard the police boat revving up. They followed us a long way while I was taking it nice and easy, cruising at about ten knots so they wouldn't lose me. They hung on to my tail for about thirty miles, waiting for the bang and laughing to themselves fit to bust, I suppose. But they didn't laugh when the bang came and blew the arse off their own boat.

"I turned and picked them up. It was all good clean fun—no one was hurt. When I'd got them out of the water I took them back to Algiers—the noble rescuer. You ought to have seen the faces of the Security boys when I pitched up. Of course, they had to go through the motions of thanking me for rescuing those lousy, shipwrecked mariners. I kept a straight face and said I thought it must have been one of the anti-submarine depth charges in the stern that had gone off. They said it couldn't have been that because police boats don't carry depth charges. And that was that."

He chuckled. "No, they don't like me in Algiers."

I laughed with him. It was a good story and he had told it well.

I was in two minds about Metcalfe; he had his advantages and his disadvantages. On the one hand, he could give us a lot of help in Tangier; he knew the ropes and had the contacts. On the other hand, we had to be careful he didn't get wind of what we were doing. He was a hell of a good chap and all that, but if he knew we were going to show up with four tons of gold he would hijack us without a second thought. We were his kind of meat.

Yes, we had to be very careful in our dealings with Mr. Metcalfe. I made a mental note to tell the others not to let anything drop in his presence.

I said, "What kind of a boat have you got?"

"A Fairmile," he said. "I've re-engined it, of course."

I knew of the Fairmiles, but I had never seen one close up. They had been built in the hundeds during the war for harbour defence. The story was that they were built by the mile and cut off as needed. They were 112 feet overall with powerfull engines and could work up over twenty knots easily, but they had the reputation of being bad rollers in a cross sea. They were not armoured or anything like that, being built of wood, and when a few of them went into St. Nazaire with the *Campbelltown* they got shot up very badly.

After the war you could buy a surplus Fairmile for about five thousand quid and they had become a favourite with the smugglers of Tangier. If Metcalfe had re-engined his Fairmile, he had probably gone for power to outrun the revenue cutters and his boat would be capable of at least twenty-six knots in an emergency. *Sanford* would have no chance of outrunning a boat like that if it came to the push.

"I'd like to see her sometime," I said. There was no harm in looking over a potential enemy.

"Sure," said Metcalfe expansively. "But not just yet. I'm going out to-morrow night."

That was good news—with Metcalfe out of the way we might be able to go about our business undisturbed. "When are you coming back?" I asked.

"Some time next week," he said. "Depending on the wind and the rain and suchlike things."

"Such as those French Security bastards?"

"That's right," he said carelessly. "Let's eat."

11

Metcalfe made us free of his flat and said we could live there in his absence—the servants would look after us. That afternoon he took me round town and introduced me to several people. Some were obviously good contacts to have, such as a ship's chandler and a boat builder. Others were not so obviously good; there was a villainous-looking café proprietor, a Greek with no discernible occupation and a Hungarian who explained volubly that he was a "Freedom Fighter" who had escaped from Hungary after the abortive revolution of 1956. I was particularly cynical about him.

I think that Metcalfe was unobtrusively passing the word that we were friends of his, and so immune to any of the usual tricks played on passing yachtsmen. Metcalfe was not a bad man to have around if he was your friend and you were a yachtsman. But I wasn't an ordinary yachtsman and that made Metcalfe a potential bomb.

Before we left the flat I had the chance to talk to Coertze and Walker privately. I said, "Here's where we keep our mouths shut and stick to our cover story. We don't do a damn' thing until Metcalfe has pushed off—and we try to finish before he gets back."

Walker said, "Why, is he dangerous?"

"Don't you know about Metcalfe?" I explained who he was. They had both heard of him; he had made quite a splash in the South African Press—the reporters loved to write about such a colourful character.

"Oh, *that* Metcalfe," said Walker, impressed.

Coertze said, "He doesn't look much to me. He won't be any trouble."

"It's not Metcalfe alone," I said. "He's got an organisation and he's on his own territory. Let's face it; he's a professional and we're amateurs. Steer clear of Metcalfe."

I felt like adding "and that's an order," but I didn't. Coertze might have taken me up on it and I didn't want to force a showdown with him yet. It would come of its own accord soon enough.

So for a day and a half we were tourists in Tangier, rubber-necking our way about the town. If we hadn't had so much on our minds it might have been interesting, but as it was, it was a waste of time.

Luckily, Metcalfe was preoccupied by his own mysterious

business and we saw little of him. However, I did instruct Walker to ask one crucial question before Metcalfe left.

Over breakfast, he said, "You know—I *like* Tangier. It might be nice to stay here for a few months. Is the climate always like this?"

"Most of the time," answered Metcalfe. "It's a good, equable climate. There's lots of people retire here, you know."

Walker smiled. "Oh, I'm not thinking of retiring. I've nothing to retire from." He was proving to be a better actor than I had expected—that touch was perfect. He said, "No, what I thought was that I might like to buy a house here. Somewhere I could live a part of the year."

"I should have thought the Med. would be your best bet," said Metcalfe. "The Riviera, or somewhere like that."

"I don't know," said Walker. "This seems to be as good a place as any, and the Riviera is *so* crowded these days." He paused as though struck by a sudden thought. "I'd want a boat, of course. Could you design one for me? I'd have it built in England."

"Sure I could," I said. "All you have to do is pay me enough."

"Yes," said Walker. "You can't do without the old boat, can you?"

He was laying it on a bit too thick and I could see that Metcalfe was regarding him with amused contempt, so I said quickly, "He's a damned good sailor. He nearly ran off with the Cape Dinghy Championship last year."

That drew Metcalfe as I knew it would. "Oh," he said with more respect, and for a few minutes he and Walker talked boats. At last Walker came out with it. "You know, what would be really perfect would be a house on the coast somewhere with its own anchorage and boat-shed. Everything self-contained, as it were."

"Thinking of joining us?" asked Metcalfe with a grin.

"Oh, no," said Walker, horrified. "I wouldn't have the nerve. I've got enough money, and besides, I don't like your smelly Fairmiles with their stinking diesel oil. No, I was thinking about a *real* boat, a sailing boat."

He turned to me. "You know, the more I think about it the better I like it. You could design a 10-tonner for me, something I could handle myself, and this place is a perfect jumping-off place for the Caribbean. A transatlantic crossing might be fun."

He confided in Metcalfe. "You know, these ocean-crossing johnnies are all very well, but most of them are broke and they have to live on their boats. Why should I do that? Think how much better it would be if I had a house here with a boat-shed at the bottom of the garden, as it were, where I could tune the boat for the trip instead of lying in that stinking harbour."

It *was* a damned good idea if you were a wealthy playboy with a yen to do a single-handed Atlantic crossing. I gave Walker full credit for his inventive powers.

Metcalfe didn't find it unreasonable, either. He said, "Not a bad idea if you can afford it. I tell you what; go and see Aristide, a friend of mine. He'll try to rent you a flat, he's got dozens empty, but tell him that I sent you and he'll be more reasonable." He scribbled an address on a piece of paper and handed it to Walker.

"Oh, thanks awfullly," said Walker. "It's really very kind of you."

Metcalfe finished his coffee. "I've got to go now; see you to-night before I leave."

When he had gone Coertze, who had sat through all this with no expression at all on his face, said, "I've been thinking about the go . . ."

I kicked his ankle and jerked my head at the Moroccan servant who had just come into the room. "*Tula,*" I said. "*Moenie hier praat nie.*" Then in English, "Let's go out and have a look round."

We left the flat and sat at a table of a nearby café. I said to Coertze, "We don't know if Metcalfe's servants speak English or not, but I'm taking no chances. Now, what did you want to say?"

He said, "I've been thinking about bringing the gold in here. How are we going to do it? You said yesterday that bullion has to be declared at Customs. We can't come in and say, 'Listen, man; I've got a golden keel on this boat and I think it weights about four tons.'"

"I've been thinking about that myself," I said. "It looks as though we'll have to smuggle it in, recast it into standard bars, smuggle it out again a few bars at a time, then bring the bars in openly and declare them at Customs."

"That's going to take time," objected Coertze. "We haven't got the time."

I sighed. "All right; let's take a good look at this time factor. To-day is 12th January and Tangier shuts up shop as far as gold is concerned on 19th April—that's—let me see, er—ninety-seven days—say fourteen weeks."

I began to calculate and to allocate the time. It would be a week before we left Tangier and another fortnight to get to

Italy. That meant another fortnight coming back, too, and I would like a week spare in case of bad weather. That disposed of six weeks. Two weeks for making preparations and for getting the gold out, and three weeks for casting the keel—eleven weeks altogether, leaving a margin of three weeks. We were cutting it fine.

I said, "We'll have to see what the score is when we get back here with the gold. Surely to God someone will buy it, even if it is in one lump. But we don't say anything until we've got it."

I began to have visions of sailing back to Egypt or even India like some sort of modern Flying Dutchman condemned to sail the seas in a million pound yacht.

Walker did not go much for these planning sessions. He was content to leave that to Coertze and me. He had been sitting listening with half an ear, studying the address which Metcalfe had given him.

Suddenly he said, "I thought old Aristide would have been an estate agent, but he's not." He read the address from the slip of paper. "'Aristide Theotopopoulis, Tangier Mercantile Bank, Boulevard Pasteur.' Maybe we could ask him something about it."

"Not a chance," I said derisively. "He's a friend of Metcalfe." I looked at Walker. "And another thing," I said. "You did very well with Metcalfe this morning, but for God's sake, don't put on that phoney Oxford accent, and less of that 'thanks awfully' stuff. Metcalfe's a hard man to fool; besides, he's been to South Africa and knows the score. You'd have done better to put on a Malmesbury accent, but it's too late to change now. But tone it down a bit, will you?"

Walker grinned and said, "O.K., old chappie."

I said, "Now we'll go and see Aristide Theoto-whatever-it-is. It wouldn't be a bad idea if we hired a car, too. It'll help us get around and it adds to the cover. We *are* supposed to be rich tourists, you know."

III

Aristide Theotopopoulis was a round man. His girth was roughly equal to his height, and as he sat down he creased in the middle like a half-inflated football bladder. Rolls of fat flowed over his collar from his jowls and the back of his neck. Even his hands were round—pudgy balls of fat with the glint of gold shining from deeply embedded rings.

"Ah, yes, Mr. Walker; you want a house," he said. "I received a phone call from Mr. Metcalfe this morning. I believe I have the very thing." His English was fluent and colloquial.

"You mean you have such a house?" inquired Walker.

"Of course! Why do you suppose Mr. Metcalfe sent you to me? He knows the Casa Saeta." He paused. "You don't mind if it's an old house?" he asked anxiously.

"Not at all," replied Walker easily. "I can afford any alterations provided the house suits me." He caught my eye, then said, hastily, "But I would like to suggest that I rent it for six months with an option to buy."

Aristide's face lengthened from a circle to an ellipse. "Very well, if that is what you wish," he said dubiously.

He took us up the north coast in a Cadillac with Coertze following in our hired car. The house looked like something from a Charles Addams' cartoon and I expected to see Boris Karloff peering from a window. There was no Moorish influence at all; it was the most hideous Victorian Gothic in the worst possible taste. But that didn't matter if it could give us what we wanted.

We went into the house and looked cursorily over the worm-eaten panelling and viewed the lack of sanitation. The kitchen was primitive and there was a shaggy garden at the back of the house. Beyond was the sea and we looked over a low cliff to the beach.

It was perfect. There was a boat-house big enough to take *Sanford* once we unstepped the mast, and there was a crude slip badly in need of repair. There was even a lean-to shed where we could set up our foundry.

I looked at everything, estimating how long it would take to put in order, then I took Coertze on one side while Aristide extolled the beauties of the house to Walker.

"What do you think?" I asked.

"Man, I think we should take it. There can't be another place like this in the whole of North Africa."

"That's just what I was thinking," I said. "I hope we can find something like this in Italy. We can get local people to fix up the slip, and with a bit of push we should be finished in a week. We'll have to do some token work on the house, but the bulk of the money must go on essentials—there'll be time to make the house livable when we come back. I'll tip Walker off about that; he's good at thinking up wacky reasons for doing the damnedest things."

We drifted back to Walker and Aristide who were still going at it hammer and tongs, and I gave Walker an imperceptible nod. He smiled dazzingly at Aristide, and said, "It's no use, Mr. Theotopopoulis, you can't talk me out of taking this house. I'm determined to have it at once—on a six months' rental, of course."

Aristide, who hadn't any intention of talking anyone out of anything, was taken aback, but making a game recovery, said, "You understand, Mr. Walker, I can give no guarantees . . ." His voice tailed off, giving the impression that he was doing Walker a favour.

"That's all right, old man," said Walker gaily. "But I must have a six months' option on the house, too. Remember that."

"I think that can be arranged," said Aristide with spurious dubiety.

"Won't it be fun, living in this beautiful house?" said Walker to me. I glared at him. That was the trouble with Walker; he got wrapped up in his part too much. My glare went unnoticed because he had turned to Aristide. "The house isn't haunted, or anything like that?" he demanded, as though he equated ghosts with dead rats in the wainscotting.

"Oh, no," said Aristide hurriedly. "No ghosts."

"A pity," said Walker negligently. "I've always wanted to live in a haunted house."

I saw Aristide changing his mind about the ghosts, so I spoke hastily to break up this buffoonery. I had no objection to Aristide thinking he was dealing with a fool, but no one could be as big a damn' fool as Walker was acting and I was afraid that Aristide might smell a rat.

I said, "Well, I suggest we go back to Mr. Theotopopoulis's office and settle the details. It's getting late and I have to do some work on the boat."

To Coertze, I said, "There's no need for you to come. We'll meet you for lunch at the restaurant we went to last night."

I had watched his blood pressure rising at Walker's fooleries and I wanted him out of the way in case he exploded. It's damned difficult working with people, especially antagonistic types like Walker and Coertze.

We went back to Aristide's office and it all went off very well. He stung us for the house, but I had no objection to that. No one who splashed money around like Walker could be anything but an honest man.

Then Walker said something that made my blood run cold,

although afterwards, on mature consideration, I conceded that he had built up his character so that he could get away with it. He said to Aristide, "Tangier is a funny place. I hear you've got bars of gold scattered about all over the place."

Aristide smiled genially. He had cut his pound of flesh and was willing to waste a few minutes in small talk; besides, this idiot Walker was going to live in Tangier—he could be milked a lot more. "Not scattered, exactly," he said. "We keep our gold in very big safes."

"Um," said Walker. "You know, it's a funny thing, but I've lived all my life in South Africa where they mine scads of gold, and I've never seen any. You can't buy gold in South Africa, you know."

Aristide raised his eyebrows as though this was unheard of.

"I've heard you can buy gold here by the pound like buying butter over the counter. It might be fun to buy some gold. Imagine me with all my money and I've never seen a gold bar," he said pathetically. "I've got a lot of money, you know. Most people say I've got too much."

Aristide frowned. This was heresy; in his book no one could have too much money. He became very earnest. "Mr. Walker, the best thing anyone can do in these troubled times is to buy gold. It's the only safe investment. The value of gold does not fluctuate like these unstable paper currencies." With a flick of his fingers he stripped the pretensions from the U.S. dollar and the pound sterling. "Gold does not rust or waste away; it is always there, always safe and valuable. If you want to invest, I am always willing to sell gold."

"Really?" said Walker. "You sell it, just like that?"

Aristide smiled. "Just like that." His smile turned to a frown. "But if you want to buy, you must buy now, because the open market in Tangier is closing very soon." He shrugged. "You say that you have never seen a bar of gold. "I'll show you bars of gold—many of them." He turned to me. "You too, Mr. Halloran, if you wish," he said off-handedly. "Please come this way."

He led us down into the bowels of the building, through grilled doors and to the front of an immense vault. On the way down, two broad-shouldered bodyguards joined us. Aristide opened the vault door, which was over two feet thick, and led us inside.

There was a lot of gold in that vault. Not four tons of it, but still a lot of gold. It was stacked up neatly in piles of bars of various sizes; it was boxed in the form of coins; it was a hell of a lot of gold.

Aristide indicated a bar. "This is a Tangier standard bar. It weighs 400 ounces troy—about twenty-seven and a half pounds avoirdupois. It is worth over five thousand pounds sterling." He picked up a smaller bar. "This is a more convenient size. It weighs a kilo—just over thirty-two ounces—and is worth about four hundred pounds.

He opened a box and let coins run lovingly through his pudgy fingers. "Here are British sovereigns—and here are American double eagles. These are French napoleons and these are Austrian ducats." He looked at Walker with a gleam in his eye and said, "You see what I mean when I say that gold never loses its value?"

He opened another box. "Not all gold coins are old. These are made privately by a bank in Tangier—not mine. This is the Tangier Hercules. It contains exactly one ounce of fine gold."

He held the coin out on the palm of his hand and let Walker take it. Walker turned it in his fingers and then passed it to me reluctantly.

It was then that this whole crazy, mad expedition ceased to be just an adventure to me. The heavy, fatty feel of that gold coin turned something in my guts and I understood what people meant when they referred to gold lust. I understood why prospectors would slave in arid, barren lands looking for gold. It is not just the value of the gold that they seek—it is gold itself. This massive, yellow metal can do something to a man; it is as much a drug as any hell-born narcotic.

My hand was trembling slightly when I handed the coin back to Aristide.

He said, tossing it, "This costs more than bullion of course, because the cost of coining must be added. But it is in a much more convenient form." He smiled sardonically. "We sell a lot to political refugees and South American dictators."

When we were back in his office, Walker said, "You have a lot of gold down there. Where do you get it from?"

Aristide shrugged. "I buy gold and I sell gold. I make my profit on both transactions. I buy it where I can; I sell it when I can. It is not illegal in Tangier."

"But it must come from somewhere," persisted Walker. "I mean, suppose one of those pirate chaps, I mean one of the smuggling fellows, came to you with half a ton of gold. Would you buy it?"

"If the price was right," said Aristide promptly.

"Without knowing where it came from?"

A faint smile came to Aristide's eyes. "There is nothing more anonymous than gold," he said. "Gold has no master; it belongs only temporarily to the man who touches it. Yes I would buy the gold."

"Even when the gold market closes?"

Aristide merely shrugged and smiled.

"Well, now, think of that," said Walker fatuously. "You must get a lot of gold coming into Tangier."

"I will sell you gold when you want it, Mr. Walker," said Aristide, seating himself behind his desk. "Now, I assume that, since you are coming to live in Tangier, you will want to open a bank account." He was suddenly all businessman.

Walker glanced at me, then said, "Well, I don't know. I'm on this cruise with Hal here, and I'm taking care of my needs with a letter of credit that was issued in South Africa. I've already cashed in a lot of boodle at one of the other banks here—I didn't realise I would have the good fortune to meet a friendly banker." He grinned engagingly.

"We're not going to stay here long," he said. "We'll be pushing off in a couple of weeks, but I'll be back; yes, I'll be back. When will we be back, Hal?"

I said, "We're going to Spain and Italy, and then to Greece. I don't think we'll push on as far as Turkey or the Lebanon, although we might. I should say we'll be back here in three or four months."

"You see," said Walker. "That's when I'll move into the house properly. Casa Saeta," he said dreamily. "That sounds fine."

We took our leave of Aristide, and when we got outside, I said furiously, "What made you do a stupid thing like that?"

"Like what?" asked Walker innocently.

"You know very well what I mean. We agreed not to mention gold."

"We've got to say something about it sometime," he said. "We can't sell gold to anyone without saying anything about it. I just thought it was a good time to find out something about it, to test Aristide's attitude towards gold of unknown origin. I thought I worked up to it rather well."

I had to give him credit for that. I said, "And another thing: let's have less of the silly ass routine. You nearly gave me a fit when you started to pull Aristide's leg about the ghosts. There are more important things at stake than fooling about."

"I know," he said soberly. "I realised that when we were in the vault. I had forgotten what gold felt like."

So it had hit him too. I calmed down and said, "O.K. But don't forget it. And for God's sake don't act the fool in front of Coertze. I have enough trouble keeping the peace as it is."

IV

When we met Coertze for lunch, I said, "We saw a hell of a lot of gold this morning."

He straightened. "Where?"

Walker said, "In a bloody big safe at Aristide's bank."

"I thought . . ." Coertze began.

"No harm done," I said. "It went very smoothly. We saw a lot of ingots. There are two standard sizes readily acceptable here in Tangier. One is 400 ounces, the other is one kilogram." Coertze frowned, and I said, "That's nearly two and a quarter pounds."

He grunted and drank his Scotch. I said, "Walker and I have been discussing this and we think that Aristide will buy the gold under the counter, even after the gold market closes—but we'll probably have to approach him before that so he can make his arrangements."

"I think we should do it now," said Walker.

I shook my head. "No! Aristide is a friend of Metcalfe; that's too much like asking a tiger to come to dinner. We mustn't tell him until we come back and then we'll have to take the chance."

Walker was silent so I went on. "The point is that it's unlikely that Aristide will relish taking a four-ton lump of gold into stock, so we'll probably have to melt the keel down into ingots, anyway. In all probability Aristide will fiddle his stock sheets somehow so that he can account for the four extra tons, but it means that he must be told before the gold market closes—which means that we must be back before April 19."

Coertze said, "Not much time."

I said, "I've worked out all the probable times for each stage of the operation and we have a month in hand. But there'll be snags and we'll need all of that. But that isn't what's worrying me now—I've got other things on my mind."

"Such as?"

"Look. When—and if—we get the gold here and we start to melt it down, we're going to have a hell of a lot of ingots lying around. I don't want to dribble them to Aristide as they're cast—that's bad policy, too much chance of an outsider catching on. I want to let him have the lot all at once, get paid with a cast-iron draft on a Swiss bank and then clear out. But it does mean that we'll have a hell of a lot of ingots lying around loose in the Casa Saeta and that's bad."

I sighed. "Where do we keep the damn' things? Stacked up in the living-room? And how many of these goddammed ingots will there be?" I added irritably.

Walker looked at Coertze. "You said there was about four tons, didn't you?"

"*Ja*," said Coertze. "But that was only an estimate."

I said, "You've worked with bullion since. How close is that estimate?"

He thought about it, sending his mind back fifteen years and comparing what he saw then with what he had learned since. The human mind is a marvellous machine. At last he said slowly, "I think it is a close estimate, very close."

"All right," I said. "So it's four tons. That's 9000 pounds as near as dammit. There's sixteen ounces to the pound and . . ."

"No," said Coertze suddenly. "Gold is measured in troy ounces. There's 14.58333 recurring ounces troy to the English pound."

He had the figures so pat that I was certain he knew what he was talking about. After all, it was his job. I said, "Let's not go into complications; let's call it fourteen and a half ounces to the pound. That's good enough."

I started to calculate, making many mistakes although it should have been a simple calculation. The mathematics of yacht design don't have the same emotional impact.

At last I had it. "As near as I can make out, in round figures we'll have about 330 bars of 400 ounces each.

"What's that at five thousand quid a bar?" asked Walker.

I scribbled on the paper again and looked at the answer unbelievingly. It was the first time I had worked this out in terms of money. Up to this time I had been too busy to think about it, and four tons of gold seemed to be a good round figure to hold in one's mind.

I said hesitantly, "I work it out as £1,650,000!"

Coertze nodded in satisfaction. "That is the figure I got. And there's the jewels on top of that."

I had my own ideas about the jewels. Aristide had been right when he said that gold is anonymous—but jewels aren't. Jewels have a personality of their own and can be traced too easily. If I had my way the jewels would stay in the tunnel. But that I had to lead up to easily.

Walker said, "That's over half a million each."

I said, "Call it half a million each, net. The odd £150,000 can go to expenses. By the time this is through we'll have spent more than we've put in the kitty."

I returned to the point at issue. "All right, we have 330 bars of gold. What do we do with them?"

Walker said meditatively. "There's a cellar in the house."

"That's a start, anyway."

He said, "You know the fantastic thought I had in that vault? I thought it looked just like a builder's yard with a lot of bricks lying all over the place. Why couldn't we build a wall in the cellar?"

I looked at Coertze and he looked at me, and we both burst out laughing.

"What's funny about that?" asked Walker plaintively.

"Nothing," I said, still spluttering. "It's perfect, that's all."

Coertze said, grinning, "I'm a fine bricklayer when the rates of pay are good."

A voice started to bleat in my ear and I turned round. It was an itinerant lottery-ticket seller poking a sheaf of tickets at me. I waved him away, but Coertze, in a good mood for once, said tolerantly, "No, man, let's have one. No harm in taking out insurance."

The ticket was a hundred pesetas, so we scraped it together from the change lying on the table, and then we went back to the flat.

V

The next day we started work in earnest. I stayed with *Sanford*, getting her ready for sea by dint of much bullying of the chandler and the sailmaker. By the end of the week I was satisfied that she was ready and was able to leave for anywhere in the world.

Coertze and Walker worked up at the house, rehabilitating the boat-shed and the slip and supervising the local labour they had found through Metcalfe's kind offices. Coertze said, "You have no trouble if you treat these wogs just the same as the Kaffirs back home." I wasn't so sure of that, but everything seemed to go all right.

By the time Metcalfe came back from whatever nefarious enterprise he had been on, we were pretty well finished and ready to leave. I said nothing to Metcalfe about this, feeling that the less he knew, the better.

When I'd got *Sanford* ship-shape I went over to Metcalfe's Fairmile to pay my promised visit. A fair-haired man who was flushing the decks with a hose said, "I guess you must be Halloran. I'm Krupke, Metcalfe's side-kick."

"Is he around?"

Krupke shook his head. "He went off with that friend of yours—Walker. He said I was to show you around if you came aboard."

I said, "You're an American, aren't you?"

He grinned. "Yep, I'm from Milwaukee. Didn't fancy going back to the States after the war, so I stayed on here. Hell, I was only a kid then, not more'n twenty, so I thought that since Uncle Sam paid my fare out here, I might as well take advantage of it."

I thought he was probably a deserter and couldn't go back to the States, although there might have been an amnesty for deserters. I didn't know how the civil statute limitations worked in military law. I didn't say anything about that, though—renegades are touchy and sometimes unaccountably patriotic.

The wheelhouse—which Krupke called the "deckhouse"—was well fitted. There were two echo sounders, one with a recording pen. Engine control was directly under the helmsman's hand and the windows in front were fitted with Kent screens for bad weather. There was a big marine radio transceiver—and there was radar.

I put my hand on the radar display and said, "What range does this have?"

"It's got several ranges," he said. "You pick the one that's best at the time. I'll show you."

He snapped a switch and turned a knob. After a few seconds the screen lit up and I could see a tiny plan of the harbour as the scanner revolved. Even *Sanford* was visible as one splotch among many.

"That's for close work," said Krupke, and turned a knob with a click. "This is maximum range—fifteen miles, but you won't see much while we're in harbour."

The landward side of the screen was now too cluttered to be of any use, but to seaward, I saw a tiny speck. "What's that?"

He looked at his watch. "That must be the ferry from Gibraltar. It's ten miles away—you can see the mileage marked on the grid."

It said, "This gadget must be handy for making a landfall at night."

"Sure," he said. "All you have to do is to match the screen profile with the chart. Doesn't matter if there's no moon or if there's a fog."

I wished I could have a set like that on *Sanford* but it's difficult installing radar on a sailing vessel—there are too many lines to catch in the antenna. Anyway, we wouldn't have the power to run it.

I looked around the wheelhouse. "With all this gear you can't need much of a crew, even though she is a biggish boat," I said. "What crew do you have?"

"Me and Metcalfe can run it ourselves," said Krupke. "Our trips aren't too long. But usually we have another man with us—that Moroccan you've got on *Sanford*.

I stayed aboard the Fairmile for a long time, but Metcalfe and Walker didn't show up, so after a while I went back to Metcalfe's flat. Coertze was already there, but there was no sign of the others, so we went to have dinner as a twosome.

Over dinner I said, "We ought to be getting away soon. Everything is fixed at this end and we'd be wasting time if we stayed any longer."

"*Ja,*" Coertze agreed. "This isn't a pleasure trip."

We went back to the flat and found it empty, apart from the servants. Coertze went to his room and I read desultorily from a magazine. About ten o'clock I heard someone coming in and looked up.

I was immediately boiling with fury.

Walker was drunk—blind, paralytic drunk. He was clutching on to Metcalfe and sagging at the knees, his face slack and his bleared eyes wavering unseeingly about him. Metcalfe was a little under the weather himself, but not too drunk. He gave Walker a hitch to prevent him from falling, and said cheerily, "We went to have a night on the town, but friend Walker couldn't take it. You'd better help me dump him on his bed."

I helped Metcalfe support Walker to his room and we laid him on his bed. Coertze, dozing in the other bed, woke up and said, "What's happening?"

Metcalfe said, "Your pal's got no head for liquor. He passed out on me."

Coertze looked at Walker, then at me, his black eyebrows drawing angrily over his eyes. I made a sign for him to keep quiet.

Metcalfe stretched and said, "Well, I think I'll turn in myself." He looked at Walker and there was an edge of contempt to his voice. "He'll be all right in the morning, barring a hell of a hangover. I'll tell Ismail to make him a prairie oyster for breakfast." He turned to Coertze. "What do you call it in Afrikaans?"

"*'n Regmaker,*" Coertze growled.

Metcalfe laughed. "That's right. A *regmaker*. That was the first word I ever learned in Afrikaans," He went to the door. "See you in the morning," he said, and was gone.

I closed the door. "The damn' fool," I said feelingly.

Coertze got out of bed and grabbed hold of Walker, shaking him. "Walker," he shouted. "Did you tell him anything?"

Walker's head flapped sideways and he began to snore. I took Coertze's shoulder. "Be quiet; you'll tell the whole household," I said. "It's no use, anyway; you won't get any sense out of him to-night—he's unconscious. Leave it till morning."

Coertze shook off my hand and turned. He had a black anger in him. "I told you," he said in a suppressed voice. "I told you he was no good. Who knows what the *dronkie* said?"

I took off Walker's shoes and covered him with a blanket. "We'll find out to-morrow," I said. "And I mean *we*. Don't you go off pop at him, you'll scare the liver out of him and he'll close up tight."

"I'll *donner* him up," said Coertze grimly. "That's God's truth."

"You'll leave him alone," I said sharply. "We may be in enough trouble without fighting among ourselves. We need Walker."

Coertze snorted.

I said, "Walker has done a job here that neither of us could have done. He has a talent for acting the damn' fool in a believable manner." I looked down at him, then said bitterly, "It's a pity he can be a damn' fool without the acting. Anyway, we may need him again, so you leave him alone. We'll both talk to him to-morrow, together."

Coertze grudgingly gave his assent and I went to my room.

VI

I was up early next morning, but not as early as Metcalfe, who had already gone out. I went in to see Walker and found that Coertze was up and half dressed. Walker lay on his bed, snoring. I took a glass of water and poured it over his head. I was in no mood to consider Walker's feelings.

He stirred and moaned and opened his eyes just as Coertze seized the carafe and emptied it over him. He sat up spluttering, then sagged back. "My head," he said, and put his hands to his temples.

Coertze seized him by the front of the shirt. "*Jou gogga-mannetjie,* what did you say to Metcalfe?" He shook Walker violently. "What did you tell him?"

This treatment was doing Walker's aching head no good, so I said, "Take it easy; I'll talk to him."

Coertze let go and I stood over Walker, waiting until he had recovered his wits. Then I said, "You got drunk last night, you stupid fool, and of all people to get drunk with you had to pick Metcalfe."

Walker looked up, the pain of his monumental hangover filming his eyes. I sat on the bed. "Now, did you tell him anything about the gold?"

"No," cried Walker. "No, I didn't."

I said evenly, "Don't tell us any lies, because if we catch you out in a lie you know what we'll do to you."

He shot a frightened glance at Coertze who was glowering in the background and closed his eyes. "I can't remember," he said. "It's a blank; I can't remember."

That was better; he was probably telling the truth now. The total blackout is a symptom of alcoholism. I thought about it for a while and came to the conclusion that even if Walker hadn't told Metcalfe about the gold he had probably blown his cover sky high. Under the influence, the character he had built up would have been irrevocably smashed and he would have reverted to his alcoholic and unpleasant self.

Metcalfe was sharp—he wouldn't have survived in his nefarious career otherwise. The change in character of Walker would be the tip-off that there was something odd about old pal Halloran and his crew. That would be enough for Metcalfe to check further. We would have to work on the assumption that Metcalfe would consider us worthy of further study.

I said, "What's done is done," and looked at Walker. His eyes were downcast and his fingers were nervously scrabbling at the edge of the blanket.

"Look at me," I said, and his eyes rose slowly to meet mine. "I think you're telling the truth," I said coldly. "But if I catch you in a lie it will be the worse for you. And if you take another drink on this trip I'll break your back. You think you're scared of Coertze here; but you'll have more reason to be scared of me if you take just one more drink. Understand?"

He nodded.

"I don't care how much you drink once this thing is finished. You'll probably drink yourself to death in six months, but that's got nothing to do with me. But just one more drink on this trip and you're a dead man."

He flinched and I turned to Coertze. "Now, leave him alone; he'll behave."

Coertze said, "Just let me get at him. Just once," he pleaded.

"It's finished," I said impatiently. "We have to decide what to do next. Get your things packed—we're moving out."

"What about Metcalfe?"

"I'll tell him we want to see some festival in Spain."

"What festival?"

"How do I know which festival? There's always some goddam festival going on in Spain; I'll pick the most convenient. We sail this afternoon as soon as I can get harbour clearance."

"I still think I could do something about Metcalfe," said Coertze meditatively.

"Leave Metcalfe alone," I said. "He *may* not suspect anything at all, but if you try to beat him up then he'll *know* there's something fishy. We don't want to tangle with Metcalfe if we can avoid it. He's bigger than we are."

We packed our bags and went to the boat, Walker very quiet and trailing in the rear. Moulay Idriss was squatting on the foredeck smoking a *kif* cigarette. We went below and started to stow our gear.

I had just pulled out the chart which covered the Straits of Gibraltar in preparation for planning our course when Coertze came aft and said in a low voice, "I think someone's been searching the boat."

"What the hell!" I said. Metcalfe *had* left very early that morning—he would have had plenty of time to give *Sanford* a good going over. "The furnaces?" I said.

We had disguised the three furnaces as well as we could. The carbon clamps had been taken off and scattered in tool boxes in the forecastle where they would look just like any other junk that accumulates over a period. The main boxes with the heavy transformers were distributed about *Sanford,* one cemented under the cabin sole, another disguised as a receiving set complete with the appropriate knobs and dials, and the third built into a marine battery in the engine space.

It is doubtful if Metcalfe would know what they were if he saw them, but the fact that they were masquerading in innocence would make him wonder a lot. It would be a certain clue that we were up to no good.

A check over the boat showed that everything was in order. Apart from the furnaces, and the spare graphite mats which lined the interior of the double coach roof, there was nothing on board to distinguish us from any other cruising yacht in these waters.

I said, "Perhaps the Moroccan has been doing some exploring on his own account."

Coertze swore. "If he's been poking his nose in where it isn't wanted I'll throw him overboard."

I went on deck. The Moroccan was still squatting on the foredeck. I said interrogatively, "Mr. Metcalfe?"

He stretched an arm and pointed across the harbour to the Fairmile. I put the dinghy over the side and rowed across. Metcalfe hailed me as I got close. "How's Walker?"

"Feeling sorry for himself," I said, as Metcalfe took the painter. "A pity it happened; he'll probably be as sick as a dog when we get under way."

"You leaving?" said Metcalfe in surprise.

I said, "I didn't get the chance to tell you last night. We're heading for Spain." I gave him my prepared story, then said, "I don't know if we'll be coming back this way. Walker will, of course, but Coertze and I might go back to South Africa by way of the east coast." I thought that there was nothing like confusing the issue.

"I'm sorry about that," said Metcalfe. "I was going to ask you to design a dinghy for me while you were here."

"Tell you what," I said. "I'll write to Cape Town and get the yard to send you a Falcon kit. It's on me; all you've got to do is pay for the shipping."

"Well, thanks," said Metcalfe. "That's decent of you." He seemed pleased.

"It's as much as I can do after all the hospitality we've had here," I said.

He stuck out his hand and I took it. "Best of luck, Hal, in all your travels. I hope your project is successful."

I was incautious. "What project?" I asked sharply.

"Why, the boatyard you're planning. You don't have anything else in mind, do you?"

I cursed myself and smiled weakly. "No, of course not."

I turned to get into the dinghy, and Metcalfe said quietly, "You're not cut out for my kind of life, Hal. Don't try it if you're thinking of it. It's tough and there's too much competition."

As I rowed back to *Sanford* I wondered if that was a veiled warning that he was on to our scheme. Metcalfe was an honest man by his rather dim lights and wouldn't willingly cut down a friend. But he would if the friend didn't get out of his way.

At three that afternoon we cleared Tangier harbour and I set course for Gibraltar. We were on our way, but we had left too many mistakes behind us.

IV. FRANCESCA

When we were beating through the Straits Coertze suggested that we should head straight for Italy. I said, "Look, we've told Metcalfe we were going to Spain, so that's where we are going."

He thumped the cockpit coaming. "But we haven't time."

"We've got to make time," I said doggedly. "I told you there would be snags which would use up our month's grace; this is one of the snags. We're going to take a month getting to Italy instead of a fortnight, which cuts us down to two weeks in hand—but we've got to do it. Maybe we can make it up in Italy."

He grumbled at that, saying I was unreasonably frightened of Metcalfe. I said, "You've waited fifteen years for this opportunity—you can afford to wait another fortnight. We're going to Gibraltar, to Malaga and Barcelona; we're going to the Riviera, to Nice and to Monte Carlo; after that, Italy. We're going to watch bullfights and gamble in casinos and do everything that every other tourist does. We're going to be the most innocent people that Metcalfe ever laid eyes on."

"But Metcalfe's back in Tangier."

I smiled thinly. "He's probably in Spain right now. He could have passed us any time in that Fairmile of his. He could even have flown or taken the ferry to Gibraltar, dammit. I think he'll keep an eye on us if he reckons we're up to something."

"Damn Walker," burst out Coertze.

"Agreed," I said. "But that's water under the bridge."

I was adding up the mistakes we had made. Number one was Walker's incautious statement to Aristide that he had drawn money on a letter of credit. That was a lie—a needless one, too—I had the letter of credit and Walker could have said so. Keeping control of the finances of the expedition was the only way I had of making sure that Coertze didn't get the jump on me. I still didn't know the location of the gold.

Now, Aristide would naturally make inquiries among his fellow bankers about the financial status of this rich Mr. Walker. He would get the information quite easily—all bankers hang together and the hell with ethics—and he would find that Mr. Walker had *not* drawn any money from any bank in Tangier. He might not be too perturbed about that, but he might ask Metcalfe about it, and Metcalfe would find it another item to add to his list of suspicions. He would pump Aristide to find that Walker and Halloran had taken an undue interest in the flow of gold in and out of Tangier.

He would go out to the Casa Saeta and sniff around. He would find nothing there to conflict with Walker's cover story, but it would be precisely the cover story that he suspected most—Walker having blown hell out of it when he was drunk. The mention of gold would set his ears a-prick—a man like Metcalfe would react very quickly to the smell of gold—and if I were Metcalfe I would take great interest in the movements of the cruising yacht, *Sanford*.

All this was predicated on the fact that Walker had *not* told about the gold when he was drunk. If he *had*, then the balloon had really gone up.

We put into Gibraltar and spent a day rubber-necking at the Barbary apes and looking at the man-made caves. Then we sailed for Malaga and heard a damn' sight more flamenco music than we could stomach.

It was on the second day in Malaga, when Walker and I went out to the gipsy caves like good tourists, that I realised we were being watched. We were bumping into a sallow young man with a moustache everywhere we went. He sat far removed when we ate in a sidewalk café, he appeared in the yacht basin, he applauded the flamenco dancers when we went to see the gipsies.

I said nothing to the others, but it only went to confirm my estimate of Metcalfe's abilities. He would have friends in every Mediterranean port, and it wouldn't be difficult to pass the word around. A yacht's movements are not easy to disguise, and he was probably sitting in Tangier like a spider in the centre of a web, receiving phone calls from wherever we went. He would know all our movements and our expenditure to the last peseta.

The only thing to do was to act the innocent and hope

that we could wear him out, string him on long enough so that he would conclude that his suspicions were unfounded, after all.

In Barcelona we went to a bull fight—the three of us. That was after I had had a little fun in trying to spot Metcalfe's man. He wasn't difficult to find if you were looking for him and turned out to be a tall, lantern-jawed cut-throat who carried out the same routine as the man in Malaga.

I was reasonably sure that if anyone was going to burgle *Sanford* it would be one of Metcalfe's friends. The word would have been passed round that we were his meat and so the lesser fry would leave us alone. I hired a watchman who looked as though he would sell his grandmother for ten pesetas and we all went to the bull-fight.

Before I left I was careful to set the stage. I had made a lot of phoney notes concerning the costs of setting up a boat-yard in Spain, together with a lot of technical stuff I had picked up. I also left a rough itinerary of our future movements as far as Greece and a list of addresses of people to be visited. I then measured to a millimetre the position in which each paper was lying.

When we got back the watchman said that all had been quiet, so I paid him off and he went away. But the papers had been moved, so the locked cabin had been successfully burgled in spite of—or probably because of—the watchman. I wondered how much he had been paid—and I wondered if my plant had satisfied Metcalfe that we were wandering innocents.

From Barcelona we struck out across the Gulf of Lions to Nice, giving Majorca a miss because time was getting short. Again I went about my business of visiting boatyards and again I spotted the watcher, but this time I made a mistake.

I told Coertze.

He boiled over. "Why didn't you tell me before?" he demanded.

"What was the point?" I said. "We can't do anything about it."

"Can't we?" he said darkly, and fell into silence.

Nothing much happened in Nice. It's a pleasant place if you haven't urgent business elsewhere, but we stayed just long enough to make our cover real and then we sailed the few miles to Monte Carlo, which again is a nice town for the visiting tourist.

In Monte Carlo I stayed aboard *Sanford* in the evening while Coertze and Walker went ashore. There was not much to do in the way of maintenance beyond the usual housekeeping jobs, so I relaxed in the cockpit enjoying the quietness of the night. The others stayed out late and when they came back Walker was unusually silent.

Coertze had gone below when I said to Walker, "What's the matter? The cat got your tongue? How did you like Monte?"

He jerked his head at the companionway. "He clobbered someone."

I went cold. "Who?"

"A chap was following us all afternoon. Coertze spotted him and said that he'd deal with it. We let this bloke follow us until it got dark and then Coertze led him into an alley and beat him up."

I got up and went below. Coertze was in the galley bathing swollen knuckles. I said, "So you've done it at last. You must use your goddamm fists and not your brains. You're worse than Walker; at least you can say he's a sick man."

Coertze looked at me in surprise. "What's the matter?"

"I hear you hit someone."

Coertze looked at his fist and grinned at me. "He'll never bother us again—he'll be in hospital for a month." He said this with pride, for God's sake.

"You've blown it," I said tightly. "I'd just about got Metcalfe to the point where he must have been convinced that we were O.K. Now you've beaten up one of his men, so he knows we are on to him, and he knows we must be hiding something. You might just as well have phoned him up and said, 'We've got some gold coming up; come and take it from us.' You're a damn' fool."

His face darkened. "No one can talk to me like that." He raised his fist.

"I *am* talking to you like that," I said. "And if you lay one finger on me you can kiss the gold good-bye. You can't sail this or any other boat worth a damn, and Walker won't help you—he hates your guts. You hit me and you're out for good. I know you could probably break me in two and you're welcome to try, but it'll cost you a cool half-million for the pleasure."

This showdown had been coming for a long time.

He hesitated uncertainly. "You damned Englishman," he said.

"Go ahead—hit me," I said, and got ready to take his rush.

He relaxed and pointed his finger at me threateningly. "You wait until this is over," he said. "Just you wait—we'll sort it out then."

"All right, we'll sort it out then," I said. "But until then I'm the boss. Understand?"

His face darkened again. "No one bosses me," he blustered.

"Right," I said. "Then we start going back the way we came—Nice, Barcelona, Malaga, Gibraltar. Walker will help me sail the boat, but he won't do a damn' thing for you." I turned away.

"Wait a minute," said Coertze and I turned back. "All right," he said hoarsely. "But wait till this is over; by God, you'll have to watch yourself then."

"But until then I'm the boss?"

"Yes," he said sullenly.

"And you take my orders?"

His fists tightened but he held himself in. "Yes."

"Then here's your first one. You don't do a damn' thing without consulting me first." I turned to go up the companion-way, got half-way up, then had a sudden thought and went below again.

I said, "And there's another thing I want to tell you. Don't get any ideas about double-crossing me or Walker, because if you do, you'll not only have me to contend with but Metcalfe as well. I'd be glad to give Metcalfe a share if you did that. And there wouldn't be a place in the world you could hide if Metcalfe got after you."

He stared at me sullenly and turned away. I went on deck.

Walker was sitting in the cockpit. "Did you hear that?" I said.

He nodded. "I'm glad you included me on your side."

I was exasperated and shaking with strain. It was no fun tangling with a bear like Coertze—he was all reflex and no brain and he could have broken me as anyone else would break a match-stick. He was a man who had to be governed like a fractious horse.

I said, "Dammit, I don't know why I came on this crazy trip with a *dronkie* like you and a maniac like Coertze. First you put Metcalfe on our tracks and then he clinches it."

Walker said softly, "I didn't mean to do it. I don't think I told Metcalfe anything."

"I don't think so either, but you gave the game away somehow." I stretched, easing my muscles. "It doesn't matter; we either get the gold or we don't. That's all there is to it."

Walker said, "You can rely on me to help you against Coertze, if it comes to that."

I smiled. Relying on Walker was like relying on a fractured mast in a hurricane—the hurricane being Coertze. He affected people like that; he had a blind, elemental force about him. An overpowering man, altogether.

I patted Walker on the knee. "O.K. You're my man from now on." I let the hardness come into my voice because Walker had to be kept to heel, too. "But keep off the booze. I meant what I said in Tangier."

II

The next stop was Rapallo, which was first choice as our Italian base, provided we could get fixed up with a suitable place to do our work. We motored into the yacht basin and damned if I didn't see a Falcon drawn up on the hard. I knew the firm had sold a few kits in Europe but I didn't expect to see any of them.

As we had come from a foreign port there were the usual Customs and medical queries—a mere formality. Yachtsmen are very well treated in the Mediterranean. I chatted with the Customs men, discussing yachts and yachting and said that I was a boat designer and builder myself. I gave the standard talk and said that I was thinking of opening a yard in the Mediterranean, pointing to the Falcon as a sample of my work.

They were impressed at that. Anyone whose product was used six thousand miles from where it was made must obviously be someone to be reckoned with. They didn't know much about local conditions but they gave me some useful addresses.

I was well satisfied. If I had to impress people with my integrity I might as well start with the Customs. That stray Falcon came in very handy.

I went ashore, leaving Walker and Coertze aboard by instruction. There was no real need for such an order but I wanted to test my new-found ascendancy over them. Coertze had returned to his old self, more or less. His mood was equable and he cracked as few jokes as usual—the point being that he cracked jokes at all. But I had no illusions that he had forgotten anything. The Afrikaner is notorious for his long memory for wrongs.

I went up to the Yacht Club and presented my credentials. One of the most pleasant things about yachting is that you are sure of a welcome in any part of the world. There is a camaraderie among yachtsmen which is very heartening in a world which is on the point of blowing itself to hell. This international brotherhood, together with the fact that the law of the sea doesn't demand a licence to operate a small boat, makes deep-sea cruising one of the most enjoyable experiences in the world.

I chatted with the secretary of the club, who spoke very good English, and talked largely of my plans. He took me into the bar and bought me a drink and introduced me to several of the members and visiting yachtsmen. After we had chatted at some length about the voyage from South Africa I got down to finding out about the local boatyards.

On the way round the Mediterranean I had come to the conclusion that my cover story need not be a cover at all—it could be the real thing. I had become phlegmatic about the gold, especially after the antics of Walker and Coertze, and my interest in the commercial possibilities of the Mediterranean was deepening. I was nervous and uncertain as to whether the three of us could carry the main job through—the three-way pull of character was causing tensions which threatened to tear the entire fabric of the plan apart. So I was hedging my bet and looking into the business possibilities seriously.

The lust for gold, which I had felt briefly in Aristide's vault, was till there but lying dormant. Still, it was enough to drive me on, enough to make me out-face Coertze and Walker and to try to circumvent Metcalfe.

But if I had known then that other interests were about to enter the field of battle I might have given up there and then, in the bar of the Rapallo Yacht Club.

During the afternoon I visited several boatyards. This was not all business prospecting—*Sanford* had come a long way and her bottom was foul. She needed taking out of the water and scraping, which would give her another half-knot. We had agreed that this would be the ostensible reason for pulling her out of the water, and a casual word dropped in the Yacht Club that I had found something wrong with her keel bolts would be enough excuse for making the exchange of keels. Therefore I was looking for a quiet place where we could cast our golden keel.

I was perturbed when I suddenly discovered that I could not spot Metcalfe's man. If he had pulled off his watchdogs because he thought we were innocent, then that was all right. But it seemed highly unlikely now that Coertze had given the game away. What seemed very likely was that something was being cooked up—and whatever was going to happen would certainly involve *Sanford*. I dropped my explorations and hurried back to the yacht basin.

"I wasn't followed," I said to Coertze.

"I told you my way was best," he said. "They've been frightened off."

"If you think that Metcalfe would be frightened off because a hired wharf rat was beaten up, you'd better think again," I said. I looked hard at him. "If you go ashore to stretch your legs can I trust you not to hammer anyone you might think is looking at you cross-eyed?"

He tried to hold my eye and then his gaze wavered. "O.K.," he said sullenly. "I'll be careful. But you'll find out that my way is best in the end."

"All right; you and Walker can go ashore to get a bite to eat." I turned to Walker. "No booze, remember. Not even wine."

Coertze said, "I'll see to that. We'll stick close together, won't we?" He clapped Walker on the back.

They climbed on to the dockside and I watched them go, Coertze striding out and Walker hurrying to keep pace. I wondered what Metcalfe was up to, but finding that profitless, I went below to review our needs for the next few days. I stretched on the port settee and must have been very tired, because when I woke it was dark except for the lights of the town glimmering through the ports.

And it was a movement on deck that had wakened me!

I lay there for a moment until I heard another sound, then I rose cautiously, went to the companionway very quietly and raised my head to deck level. "Coertze?" I called softly.

A voice said, "Is that Signor Halloran?" The voice was very feminine.

I came up into the cockpit fast. "Who is that?"

A dark shape moved towards me. "Mr. Halloran, I want to talk to you." She spoke good English with but a trace of Italian accent and her voice was pleasantly low and even.

I said, "Who are you?"

"Surely introductions would be more in order if we could see each other." There was a hint of command in her voice as though she was accustomed to getting her own way.

"O.K.," I said. "Let's go below."

She slipped past me and went down the companionway and I followed, switching on the main cabin lights. She turned so that I could see her, and she was something worth looking at. Her hair was raven black and swept up into smooth wings on each side of her head as though to match the winged eyebrows which were dark over cool, hazel eyes. Her cheekbones were high, giving a trace of hollow in the cheeks, but she didn't look like one of the fashionably emaciated models one sees in *Vogue*.

She was dressed in a simple woollen sheath which showed off a good figure to perfection. It might have been bought at a local department store or it might have come from a Parisian fashion house; I judged the latter—you can't be married to a woman for long without becoming aware of the price of feminine fripperies.

She carried her shoes in her hand and stood in her stockinged feet, that was a point in her favour. A hundred-pound girl in a spike heel comes down with a force of two tons, and that's hell on deck planking. She either knew something about yachts or . . .

I pointed to the shoes and said, "You're a pretty inexperienced burglar. You ought to have those slung round your neck to leave your hands free."

She laughed. "I'm not a burglar, Mr. Halloran, I just don't like shoes very much; and I have been on yachts before."

I moved towards her. She was tall, almost as tall as myself. I judged her to be in her late twenties or possibly, but improbably, her early thirties. Her lips were pale and she wore very little make-up. She was a very beautiful woman.

"You have the advantage of me," I said.

"I am the Contessa di Estrenoli."

I gestured at the settee. "Well, sit down, Contessa."

"Not Contessa—Madame," she said, and sat down, pulling the dress over her knees with one hand and placing the shoes at her side. "In our association together you will call me Madame."

I sat down slowly on the opposite settee. Metcalfe certainly came up with some surprises. I said carefully, "So we are going to be associated together? I couldn't think of a better person to be associated with. When do we start?"

There was frost in her voice. "Not the kind of association you are obviously thinking of, Mr. Halloran." She went off at a tangent. "I saw your . . . er . . . companions ashore. They didn't see me—I wanted to talk to you alone."

"We're alone," I said briefly.

She gathered her thoughts, then said precisely, "Mr. Halloran, you have come to Italy with Mr. Coertze and Mr. Walker to remove something valuable from the country. You intend to do this illicitly and illegally, therefore your whole plan depends on secrecy; you cannot—shall we say 'operate'

—if someone is looking over your shoulder. I intend to look over your shoulder."

I groaned mentally. Metcalfe had the whole story. Apparently the only thing he didn't know was where the treasure was hidden. This girl was quite right when she said that it couldn't be lifted if we were under observation, so he was coming right out and asking for a cut. Walker really *must* have talked in Tangier if Metcalfe could pinpoint it as close as Rapallo.

I said, "O.K., Contessa; how much does Metcalfe want?"

She raised her winged eyebrows. "Metcalfe?"

"Yes, Metcalfe; your boss."

She shook her head. "I know of no Metcalfe, whoever he is. And I am my own boss, I assure you of that."

I think I kept my face straight. The surprises were certainly piling up. If this Estrenoli woman was mixed up with Metcalfe, then why would she deny it? If she wasn't, then who the devil was she—and how did she know of the treasure?

I said, "Supposing I tell you to jump over the side?"

She smiled. "Then you will never get these valuables out of Italy."

There seemed to be a concession there, so I said, "And if I *don't* tell you to jump over the side, then we *will* get the stuff out of the country, is that it?"

"Some of it," she compromised. "But without my co-operation you will spend a long time in an Italian prison."

That was certainly something to think about and when I had time. I said, "All right; who are you, and what do you know?"

"I knew that the news was out on the waterfront to watch for the yacht *Sanford*. I knew that the yacht was owned by Mr. Halloran and that Mr. Coertze and Mr. Walker were his companions. That was enough for me."

"And what has the Contessa di Estrenoli got to do with waterfront rumours? What has an Italian aristocrat got to do with the jailbirds that news was intended for?"

She smiled and said, "I have strange friends, Mr. Halloran. I learn all that is interesting on the waterfront. I realise now that perhaps your Mr. Metcalfe was responsible for the circulation of those instructions."

"So you learned that a yacht and three men were coming to Rapallo, and you said to yourself, 'Ah, these three men are coming to take something out of Italy illegally,'" I said with heavy irony. "You'll have to do better than that, Contessa."

"But you see, I know Mr. Coertze and Mr. Walker," she said. "The heavy and clumsy Mr. Coertze has been to Italy quite often. I have always known about him and I always had him watched." She smiled. "He was like a dog at a rabbit hole who yelps because it is too small and he cannot get in. He always left Italy empty-handed."

That did it. Coertze must have shown his hand on one of his periodic trips to Italy. But how the devil did she know Walker? He hadn't been to Italy recently—or had he?

She continued. "So when I heard that Mr. Coertze was returning with Mr. Walker and the unknown Mr. Halloran, then I knew that something big was going to happen. That you were ready to take away whatever was buried, Mr. Halloran."

"So you don't know exactly what we're after?"

"I know that it is very valuable," she said simply.

"I might be an archæologist," I said.

She laughed. "No, you are not an archæologist, Mr. Halloran; you are a boat-builder." She saw the surprise in my eyes, and added, "I know a lot about you."

I said, "Let's quit fencing; how do you know about whatever it is?"

She said slowly, "A man called Alberto Corso had been writing a letter to my father. He was killed before the letter was finished, so there was not all the information that could be desired. But there was enough for me to know that Mr. Coertze must be watched."

I snapped my fingers. "You're the Count's little daughter. You're . . . er . . . Francesca."

She inclined her head. "I am the daughter of a count."

"Not so little now," I said. "So the Count is after the loot."

Her eyes widened. "Oh, no. My father knows nothing about it. Nothing at all."

I thought that could do with a bit of explanation and was just going to query the statement when someone jumped on deck. "Who is that?" asked the Contessa.

"Probably the others coming back," I said, and waited. Perhaps there were to be some more surprises before the evening was out.

Walker came down the companionway and stopped when he saw her. "Oh," he said. "I hope I'm not butting in."

I said, "This is the Contessa di Estrenoli—Mr. Walker."

I watched him to see if he recognised her, but he didn't. He looked at her as one looks at a beautiful woman and said, in Italian, " A pleasure, signora."

She smiled at him and said, " Don't you know me, Mr. Walker? I bandaged your leg when you were brought into the hill camp during the war."

He looked at her closely and said incredulously, " Francesca!"

" That's right; I'm Francesca."

" You've changed," he said. " You've grown up. I mean . . . er . . ." he was confused.

She looked at him. " Yes, we've all changed," she said. I thought I detected a note of regret. They chatted for a few minutes and then she picked up her shoes. " I must go," she said.

Walker said, " But you've only just got here."

" No, I have an appointment in twenty minutes." She rose and went to the companionway and I escorted her on deck.

She said, " I can understand Coertze, and now I can understand Walker; but I connot understand you, Mr. Halloran. Why are you doing this? You are a successful man, you have made a name in an honourable profession. Why should you do this?"

I sighed and said, " I had a reason in the beginning; maybe I still have it—I don't know. But having come this far I must go on."

She nodded, then said, " There is a café on the waterfront called the Three Fishes. Meet me there at nine to-morrow morning. Come alone; don't bring Coertze or Walker. I never liked Coertze, and now I don't think I like Walker any more. I would prefer not to talk to them."

" All right," I said. " I'll be there."

She jumped lightly on to the jetty and swayed a little as she put her shoes on. I watched her go away, hearing the sharp click of her heels long after the darkness had swallowed her. Then I went below.

Walker said, " Where did she come from? How did she know we were here?"

" The gaff has been blown with a loud trumpeting noise," I said. " She knows all—or practically all—and she's putting the screws on."

Walker's jaw dropped. " She knows about the gold?"

" Yes," I said. " But I'm not going to talk about it till Coertze comes. No point in going over it twice."

Walker protested, but swallowed his impatience when I

made it clear that I wasn't going to talk, and sat wriggling on the settee. After half an hour we heard Coertze come on board.

He was affable—full of someone else's cooking for a change, and he'd had a few drinks. "Man," he said, "these Italians can cook."

"Francesca was here," I said.

He looked at me, startled. "The Count's daughter?"

"Yes."

Walker said, "I want to know how she found us."

"What did the stuck-up bitch want?" asked Coertze.

I raised my eyebrows at that. Apparently the dislike between these two was mutual. "She wants a cut of the treasure," I said bluntly.

Coertze swore. "How the hell did she get to know about it?"

"Alberto wrote a letter before he was killed."

Coertze and Walker exchanged looks, and after a pregnant silence, Coertze said, "So Alberto was going to give us away, after all."

I said, "He *did* give you away."

"Then why is the gold still there?" demanded Coertze.

"The letter was incomplete," I said. "It didn't say exactly where the gold is."

Coertze sighed windily. "Well, there's not too much damage done."

I fretted at his stupidity. "How do you suppose we're going to get it out with half of Italy watching us?" I asked. "She's been on to you all the time—she's watched you every time you've been in Italy and she's been laughing at you. And she knows there's something big under way now."

"That bitch would laugh at me," said Coertze viciously. "She always treated me like dirt. I suppose the Count has been laughing like hell, too."

I rubbed my chin thoughtfully. "She says the Count knows nothing about it. Tell me about him."

"The Count? Oh, he's an old no-good now. He didn't get his estates back after the war—I don't know why—and he's as poor as a church mouse. He lives in a poky flat in Milan with hardly enough room to swing a cat."

"Who supports him?"

Coertze shrugged. "I dunno. Maybe she does—she can afford it. She married a Roman count; I heard he was stinking rich, so I suppose she passes on some of the housekeeping money to the old boy."

"Why don't you like her?"

"Oh, she's one of these stuck-up society bitches—I never did like that kind. We get plenty in Houghton, but they're worse here. She wouldn't give me the time of day. Not like her old man. I get on well with him."

I thought perhaps that on one of his visits to Italy Coertze had made a pass at her and been well and truly slapped down. A pass from Coertze would be clumsy and graceless, like being propositioned by a gorilla.

I said, "Was she around often during the times you were in Italy?"

He thought about that, and said, "Sometimes. She turned up at least once on every trip."

"That's all she'd need. To locate you, I mean. She seems to have a circle of pretty useful friends and apparently they're not the crowd you'd think a girl like that would mix with. She picked up Metcalfe's signals to the Mediterranean ports and interpreted them correctly, so it looks as though she has brains as well as beauty."

Coertze snorted. "Beauty! She's a skinny bitch."

She *had* got under his skin. I said, "That may be, but she's got us cold. We can't do a damn' thing while she's on our necks. To say nothing of Metcalfe, who'll be on to us next. Funny that he hasn't shown his hand in Rapallo yet."

"I tell you he's scared off," growled Coertze.

I let that pass. "Anyway, we can't do any heavy thinking about it until we find out exactly what she wants. I'm seeing her to-morrow morning, so perhaps I'll be able to tell you more after that."

"I'll come with you," said Coertze instantly.

"She wants to see me, not you," I said. "That was something she specified."

"The bloody little bitch," exploded Coertze.

"And for God's sake, think up another word; I'm tired of that one," I said irritably.

He glowered at me. "You falling for her?"

I said wearily, "I don't know the woman—I've seen her for just fifteen minutes. I'll be better able to tell you about that to-morrow, too."

"Did she say anything about me?" asked Walker.

"No," I lied. There wasn't any point in having both of them irritated at her—it was likely that we'd all have to work closely together, and the less friction the better. "But I'd better see her alone."

Coertze growled under his breath, and I said, "Don't worry;

neither she nor I know where the gold is. We still need you —she and I and Metcalfe. We mustn't forget Metcalfe."

III

Early next morning I went to find the Three Fishes. It was just an ordinary dockside café, the kind of dump you find on any waterfront. Having marked it, I went for a stroll round the yacht basin, looking at the sleek sailing yachts and motor craft of the European rich. A lot were big boats needing a paid crew to handle them while the owner and his guests took it easy, but some were more to my taste—small, handy sailing cruisers run by their owners who weren't afraid of a bit of work.

After a pleasant hour I began to feel hungry so I went back to the Three Fishes for a late breakfast and got there on the dot of nine. She wasn't there, so I ordered breakfast and it turned out better than I expected. I had just started to eat when she slid into the seat opposite.

"Sorry I'm late," she said.

"That's O.K."

She was wearing slacks and sweater, the kind of clothes you see in the women's magazines but seldom in real life. The sweater suited her.

She looked at my plate and said, "I had an early breakfast, but I think I'll have another. Do you mind if I join you?"

"It's your party."

"The food is good here," she said, and called a waiter, ordering in rapid Italian. I continued to eat and said nothing. It was up to her to make the first move. As I had said—it was her party.

She didn't say anything, either; but just watched me eat. When her own breakfast arrived she attacked it as though she hadn't eaten for a week. She was a healthy girl with a healthy appetite. I finished my breakfast and produced a packet of cigarettes. "Do you mind?" I asked.

I caught her with her mouth full and she shook her head, so I lit a cigarette. At last she pushed her plate aside with a sigh and took the cigarette I offered. "Do you know our *Espresso*?" she asked.

"Yes, I know it."

She laughed. "Oh, yes, I forgot that it must have penetrated even your Darkest Africa. It is supposed to be for after dinner, but I drink it all the time. Would you like some?"

I said that I would, so she called out to the waiter, "*Due*

Espressi," and turned back to me. Well, Mr. Halloran, have you thought about our conversation of last night?"

I said I had thought about it.

"And so?"

"And so," I repeated. "Or more precisely—so what? I'll need to know a lot more about you before I start confiding in you, Contessa."

She seemed put out. "Don't call me Contessa," she said pettishly. "What do you want to know?"

I flicked ash into the ash-tray. "For one thing, how did you intercept Metcalfe's message? It doesn't seem a likely thing for a Contessa to come across—just like that."

"I told you I have friends," she said coldly.

"Who are these friends?"

She sighed. "You know that my father and I were rebels against the Fascist Government during the war?"

"You were with the partisans, I know."

She gestured with her hand. "All right, with the partisans, if you wish. Although do not let my friends hear you say that—the Communists have made it a dirty word. My friends were also partisans and I have never lost contact with them. You see, I was only a little girl at the time and they made me a sort of mascot of the brigade. After the war most of them went back to their work, but some of them had never known any sort of life other than killing Germans. It is a hard thing to forget, you understand?"

I said, "You mean they'd had a taste of adventure, and liked it."

"That is right. There was plenty of adventure even after the war. Some of them stopped killing Germans and started to kill Communists—Italian Communists. It was dreadful. But the Communists were too strong, anyway. A few turned to other adventures—some are criminals—nothing serious, you understand; some smuggling, some things worse, but nothing very terrible in most cases. Being criminals, they also know other criminals."

I began to see how it had been worked; it was all very logical, really.

"There is a big man in Genoa, Torloni; he is a leader of criminals, a very big man in that sort of thing. He sent word to Savona, to Livorno, to Rapallo, to places as far south as Napoli, that he was interested in you and would pay for any information. He gave all your names and the name of your boat."

That was the sort of pull Metcalfe would have. Probably this Torloni owed him a favour and was paying it off.

Francesca said, " My friends heard the name—Coertze. It is very uncommon in Italy, and they knew I was interested in a man of that name, so I was told of this. When I also heard the name of Walker I was sure that something was happening." She shrugged. " And then there was this Halloran—you. I did not know about you, so I am finding out."

" Has Torloni been told about us?"

She shook her head. " I told my friends to see that Torloni was not told. My friends are very strong on this coast; during the war all these hills belonged to us—not to the Germans."

I began to get the picture. Francesca had been the mascot and, besides, she was the daughter of the revered leader. She was the Lady of the Manor, the Young Mistress who could do no wrong.

It looked also as though, just by chance, Metcalfe had been stymied—temporarily, at least. But I was landed with Francesca and her gang of merry men who had the advantage of knowing just what they wanted.

I said, " There's another thing. You said your father doesn't know anything about this. How can that be when Alberto Corso wrote him a letter?"

" I never gave it to him," she said simply.

I looked at her quizzically. " Is that how a daughter behaves to her father? Not only reading his correspondence, but withholding it as well."

" It was not like that at all," she said sharply. " I will tell you how it was." She leaned her elbows on the table. " I was very young during the war, but my father made me work, everyone had to work. It was one of my tasks to gather together the possessions of those who were killed so that useful things could be saved and anything personal could be passed on to the family.

" When Alberto was killed on the cliff I gathered his few things and I found the letter. It was addressed to my father and there were two pages, otherwise it was unfinished. I read it briefly and it seemed important, but how important it was I did not know because I was very young. I put it in my pocket to give to my father.

" But there was a German attack and we had to move. We sheltered in a farmhouse but we had to move even from there very quickly. Now, I carried my own possessions in a little tin box and that was left in the farmhouse. It was only in

1946 that I went back to the farm to thank those people—the first chance I had.

"They gave me wine and then the farmer's wife brought out the little box and asked if it was mine. I had forgotten all about it and I had forgotten what was in it." She smiled. "There was a doll—no, not a doll; what you call an . . . Eddy-bear?"

"A Teddy-bear."

"That is right; a Teddy-bear—I have still got it. There were some other things and Alberto's letter was there also."

I said, "And you still didn't give it to your father. Why not?"

She thumped the table with a small fist. "It is difficult for you to understand the Italy of just after the war, but I will try to explain. The Communists were very strong, especially here in the north, and they ruined my father after the war. They said he had been a collaborationist and that he had fought the Communist partisans instead of fighting the Fascists. My father, who had been fighting the Fascists all his life! They brought up false evidence and no one would listen to him.

"His estates had been confiscated by the Fascist Government and he could not get them back. How could he when Togliatti, the Vice-Premier of the Government, was the leader of the Italian Communist Party? They said, 'No, this man was a collaborator, so he must be punished.' But even with all their false evidence they dared not bring him to trial, but he could not get back his estates, and to-day he is a poor man."

Francesca's eyes were full of tears. She wiped them with a tissue and said, "Excuse me, but my feeling on this is strong."

I said awkwardly, "That's all right."

She looked up and said, "These Communists with their fighting against the Fascists. My father fought ten times harder than any of them. Have you heard of the 52nd Partisan Brigade?"

I shook my head.

"That was the famous Communist Brigade which captured Mussolini. The famous Garibaldi Brigade. Do you know how many men were in this so-famous Garibaldi Brigade in 1945?"

I said, "I know very little about it."

"Eighteen men," she said contemptuously. "Eighteen men called themselves the 52nd Brigade. My father commanded fifty times as many men. But when I went to Parma for the anniversary celebrations in 1949 the Garibaldi Brigade marched through the street and there were hundreds of men. All the Communist scum had crawled out of their holes now the war was over and it was safe. They marched through the streets and every man wore a red scarf about his neck and every man called himself a partisan. They even painted the statue of Garibaldi so that it had a red shirt and a red hat. So my friends and I do not call ourselves partisans, and you must not call us by that word the Communists have made a mockery of."

She was shaking with rage. Her fists were clenched and she looked at me with eyes bright with unshed tears.

"The Communists ruined my father because they knew he was a strong man and because they knew he would oppose them in Italy. He was a liberal, he was for the middle of the road—the middle way. He who is in the middle of the road gets knocked down, but he could not understand that," she said sombrely. "He thought it was an honourable fight—as though the Communists have ever fought honourably."

It was a moving story and typical of our times. I also observed that it fitted in with what Coertze had told me. I said, "But the Communists are not nearly as strong to-day. Is it not possible for your father to appeal and to have his case reviewed?'

"Mud sticks, whoever throws it," she said sadly. "Besides, the war was a long time ago—people do not like to be reminded about those times—and people, especially officials, never like to admit their mistakes."

She was realistic about the world and I realised that I must be realistic too. I said, "But what has this got to do with the letter?"

"You wanted to know why I did not give the letter to my father after the war; is that not so?"

"Yes."

She smiled tightly. "You must meet my father and then you would understand. You see, whatever you are looking for is valuable. I understood from Alberto's letter that there are papers and a lot of gold bars. Now, my father is an honourable man. He would return everything to the Government because from the Government it came. To him, it would be unthinkable to keep any of the gold for himself. It would be dishonourable."

She looked down at the backs of her hands. "Now, I am not an honourable woman. It hurts me to see my father so poor he has to live in a Milan slum, that he has to sell his furniture to buy food to eat. He is an old man—it is not right that he should live like that. But if I can get some money I would see that he had a happy old age. He does not need to know where the money comes from."

I leaned back in my chair and looked at her thoughtfully. I looked at the expensive, fashion-plate clothing she was wearing, and she coloured under my scrutiny. I said softly, "Why don't *you* send him money. I hear you made a good marriage; you ought to be able to spare a little for an old man."

Her lips twisted in a harsh smile. "You don't know anything about me, do you, Mr. Halloran? I can assure you that I have no money and no husband, either—or no one that I would care to call my husband." She moved her hands forward on the table. "I sold my rings to get money to send to my father, and that was a long time ago. If it were not for my friends I would be on the streets. No, I have no money, Mr. Halloran."

There was something here I did not understand, but I didn't press it. The reason she wanted to cut in didn't matter; all that mattered was she had us over a barrel. With her connections we could not make a move in Italy without falling over an ex-partisan friend of hers. If we tried to lift the gold without coming to terms with her she would coolly step in at the right time and take the lot. She had us taped.

I said, "You're as bad as Metcalfe."

"That is something I wanted to ask you," she said. "Who is this Metcalfe?"

"He's up to the same lark that you are."

Her command of English was not up to that. "Lark?" she said in mystification. "That is a bird?"

I said, "He's one of our mutual competitors. He's after the gold, too." I leaned over the table. "Now, if we cut you in, we would want certain guarantees."

"I do not think you are in a position to demand guarantees," she said coldly.

"Nevertheless, we would want them. Don't worry, this is in your interest, too. Metcalfe is the man behind Torloni and he's quite a boy. Now, we would want protection against Metcalfe and anything he could throw against us. From what you've said, Torloni carries a bit of weight, and if he hasn't got enough, Metcalfe can probably drum up some more. What I want to know is—can you give us protection against that lot?"

"I can find a hundred men, any time I want," she said proudly.

"What kind of men?" I asked bluntly. "Old soldiers on pension?"

She smiled. "Most of my war-time friends live quietly and go about their work. I would not want them to be mixed up in anything illegal or violent, although they would help if they had to. But my . . ." she hesitated for a word, ". . . . my more unsavoury friends I would willingly commit to this affair. I told you they are adventurous and they are not old men—no older than you, Mr. Halloran," she ended sweetly.

"A hundred of them?"

She thought a little. "Fifty, then," she compromised. "My father's hill fighters will be more than a match for those dock-land gangsters."

I had no doubt about that—if they fought man to man. But Metcalfe and Torloni could probably whip up every thug in Italy, and would do for a stake as large as this.

I said, "I want further guarantees. How do I know we won't be double-crossed?"

"You don't," she said meagerly.

I decided to go in for some melodramatics. "I want you to swear that you won't double-cross us."

She raised her hand. "I swear that I, Francesca di Estrenoli, promise faithfully not to trick, in any way, Mr. Halloran of South Africa." She smiled at me. "Is that good enough?"

I shook my head. "No, it isn't enough. You said yourself that you were a dishonourable woman. No, I want you to swear on your father's name and honour."

Pink anger spots burned on her cheeks and I thought for a moment that she was going to slap my face. I said gently, "Do you swear?"

She dropped her eyes. "I swear," she said in a low voice.

"On your father's name and honour," I persisted.

"On my father's name and on his honour," she said, and looked up. "Now I hope you are satisfied." There were tears in her eyes again.

I relaxed. It wasn't much but it was the best I could do and I hoped it would hold her.

The man from behind the counter came over to the table slowly. He looked at me with dislike and said to Francesca, "Is everything all right, madame?"

"Yes, Giuseppi, everything is all right." She smiled at him. "Nothing is wrong."

Giuseppi smiled back at her, gave me a hard look and returned to the counter. I felt a prickle at the back of my neck.

I had the feeling that if Francesca had said that everything was *not* all right I would have been a candidate for a watery dockside grave before the week was out.

I cocked my thumb at the counter. "One of your soldier friends?"

She nodded. "He saw you had hurt me, so he came over to see what he could do."

"I didn't mean to hurt you," I said.

"You shouldn't have come here. You shouldn't have come to Italy. What is it to you? I can understand Coertze and Walker; they fought the Germans, they buried the gold. But I cannot understand you."

I said gently, "I fought the Germans, too, in Holland and Germany."

"I'm sorry," she said. "I shouldn't have said that."

"That's all right. As for the rest . . ." I shrugged. "Somebody had to plan—Coertze and Walker couldn't do it. Walker is an alcoholic and Coertze is all beef and no subtlety. They needed someone to get behind and push."

"But why is it you who has to push?"

"I had a reason once," I said shortly. "Forget it. Let's get some things straightened out. What about the split?"

"The split?"

"How do we divide the loot?"

"I hadn't thought of that—it will need some thinking about."

"It will," I agreed. "Now, there's the three of us, there's you and there's fifty of your friends—fifty-four in all. If you're thinking along the lines of fifty-four equal shares you can forget about it. We won't have it."

"I can't see how we can work this out when we don't know how much money will be involved."

"We work it on a percentage basis," I said impatiently. "This is how I see it—one share each for the three of us, one share for you and one share to be divided among your friends."

"No," she said firmly. "That's not fair. You have done nothing about this, at all. You are just a plunderer."

"I thought you'd take that attitude," I said. "Now, listen, and listen damned carefully because I'm not going to repeat this. Coertze and Walker are entitled to a share each. They fought for the gold and they disposed of it carefully. Besides, they are the only people who know where it is. Right?"

She nodded agreement.

I smiled grimly. "Now we come to me whom you seem to despise." She made a sudden gesture with her hand and

I waved her down. "I'm the brains behind this. I know a way of getting the stuff out of Italy and I've arranged a sale for it. Without me this whole plan would flop, and I've invested a lot of time and money in it. Therefore I think I'm entitled to an equal share."

I stabbed my finger at her. "And now you come along and blackmail us. Yes, blackmail," I said as she opened her mouth to protest. "You've done nothing constructive towards the plan and you complain about getting an equal share. As for your friends, as far as I'm concerned, they are hired muscle to be paid for. If you don't think they're being paid enough with one-fifth between them you can supplement it out of your own share."

"But it will be so little for them," she said.

"Little!" I said, and was shocked into speechlessness. I recovered my breath. "Do you know how much is involved?"

"Not exactly," she said cautiously.

I threw discretion to the winds. "There's over £1,500,000 in gold alone—and there's probably an equal amount in cut gem-stones. The gold alone means £300,000 for a fifth share and that's £6,000 each for your friends. If you count the jewels you can double those figures."

Her eyes widened as she mentally computed this into lire. It was an astronomical calculation and took her some time. "So much," she whispered.

"So much," I said. I had just had an idea. The gems had been worrying me because they would be hot—in the criminal sense. They would need recutting and disguising and the whole thing would be risky. Now I saw the chance of doing an advantageous deal.

"Look here," I said generously. "I've just offered you and your friends two-fifths of the take. Supposing the jewels are worth more than two-fifths—and I reckon they are—then you can take the lot of them, leaving the disposal of the gold to the three of us. After all, gems are more portable and easily hidden."

She fell for it. "I know a jeweller who was with us during the war; he could do the valuation. Yes, that seems reasonable."

It seemed reasonable to me, also, since I had been taking only the gold into my calculations all the time. Coertze, Walker and myself would still come out with half a million each.

"There's one other thing," I said.

"What's that?"

"There's a lot of paper money in this hoard—lire, francs,

dollars and so on. Nobody takes any of that—there'll be records of the numbers lodged with every bank in the world. You'll have to control your friends when it comes to that."

"I can control them," she said loftily. She smiled and held out her hand. "It's a deal, then, as the Americans say."

I looked at her hand but didn't touch it. I shook my head. "Not yet. I still have to discuss it with Coertze and Walker. They'll take a hell of a lot of convincing—especially Coertze. What did you do to him, anyway?"

She withdrew her hand slowly and looked at me strangely. "Almost you convince me that you are an honest man."

I grinned at her cheerfully. "Out of necessity, that's all. Those two are the only ones who know where the gold is."

"Oh, yes, I had forgotten. As for Coertze, he is a boor."

"He'd be the first to agree with you," I said. "But it means something different in Afrikaans." I had a sudden thought. "Does anyone else know what you know—about Alberto's letter and all that?"

She started to shake her head but stopped suddenly, deciding to be honest. "Yes," she said. "One man, but he can be trusted—he is a true friend."

"O.K.," I said. "I just wanted to be sure that no one else will try to pull the same stunt that you've just pulled. The whole damn' Mediterranean seems to be getting into the act. I wouldn't tell your friends anything you don't have to—at least, not until it's all over. If they are criminals, as you say, they might get their own ideas."

"I haven't told them anything so far, and I'm not going to tell them now."

"Good. But you *can* tell them to watch for Torloni's men. They'll be keeping an eye on *Sanford* when they get round to finding where she is."

"Oh, yes, Mr. Halloran; I'll certainly tell them to keep a watch on your boat," she said sweetly.

I laughed. "I know you will. When you've got things organised drop in and see us anytime—but make it quick, there's a time limit on all this."

I got up from the table and left her. I thought she might as well pay for the breakfast since we were partners—or, as she had put it, "in association."

IV

She came that afternoon, accompanied by a man even bigger than Coertze, whom she introduced as Piero Morese.. He nodded civilly enough to me, ignored Walker and regarded Coertze watchfully.

I had had trouble with Coertze—he had taken a lot of convincing and had reiterated in a bass growl, "I will not be cheated, I will not be cheated."

I said wearily, "O.K. The gold is up in those hills somewhere; you know where it is. Why don't you go and get it? I'm sure you can fight Torloni and Metcalfe and the Contessa and her cut-throats single-handed; I'm sure you can bring back the gold and take it to Tangier before April 19. Why don't you just go ahead and stop bothering me?"

He had calmed down but was not altogether happy and he rumbled like a volcano which does not know whether to erupt again or not. Now he sat in the cabin looking at the Contessa with contempt and the big Italian with mistrust.

Morese had no English so the meeting came to order in Italian, which I could understand if it was not spoken too quickly. The Contessa said, "It is all right to speak in front of Piero, he knows everything that I know."

"I know you: you were with Umberto," said Coertze in mashed Italian.

Morese gave a quick nod but said nothing. The Contessa said, "Here is where we talk seriously." She looked at me. "Have you talked this over?"

"We have."

"Do they accept the terms?"

"They do."

"Very well, where is the gold?"

There was a growl from Coertze which I covered with a quick burst of laughter. "Contessa, you'll be the death of me," I said. "I'll die laughing. You don't suppose we'll tell you that, do you?"

She smiled acidly. "No—but I thought I would try it. All right, how do we go about this?"

I said, "First of all, there's a time limit. We'll want the gold delivered to Rapallo by the 1st of March at the latest. We also want a place where we can work undisturbed with this boat; either a private boat-shed or a boatyard. That must be arranged for now."

Her eyes narrowed. "Why the 1st of March?"

"That is of no consequence to you, but that is the way it must be."

Morese said, "That does not leave us much time. The first of the month is in two weeks."

"True," I said. "But that is the way it must be. The next thing is that only the five of us here will go to the gold. There must be no one else. We will unseal the place where it is hidden, pack what we want into strong boxes and move it out. Then we will seal the hidden place again. After that, and only after that, will we need the help of anyone else, and even then, only for lifting and transport to the coast. There is no need to have too many people knowing what we are doing."

"That is well thought of," said Morese.

I said, "Everything will be brought to the boat-shed—everything, including the jewels. We five will live together for one month while my friends and I do what we have to do. If you want the jewels valued you must bring your valuer to the jewels—not vice versa. The final share-out will be decided when the stones have been valued, but will not take place until the boat is in the water."

"You talk as though you do not trust us," said Morese.

"I don't," I said bluntly. I jerked my thumb at the Contessa. "You're friend here is blackmailing us into all this, so I don't see where the trust comes in."

His face darkened. "That is unworthy of you."

I shrugged. "Say, rather, it is unworthy of her. She started all this and those are the facts."

The Contessa put her hand on Morese's shoulder and he subsided. Coertze barked a short laugh. "*Magtig,* but you have taken her measure." He nodded. "You'll have to watch her, she's a *slim meisie.*"

I turned to him. "Now it's up to you. What will you need to get the gold?"

Coertze leaned forward. "When I was here last year nothing had changed or been disturbed. The place is in the hills where no one goes. There is a rough road so we can take a lorry right up to the place. The nearest village is four miles away."

"Can we work at night?" I asked.

Coertze thought about that. "The fall of rock looks worse than it is," he said. "I know how to blast and I made sure of that. Two men with picks and shovels will be able to get through in four hours—longer at night, perhaps—I would say six hours at night."

"So we will be there at least one whole night and probably longer."

"*Ja*," he said. "If we work at night only, it will take two nights."

The Contessa said, "Italians do not walk the hills at night. It will be safe to have lights if they cannot be seen from the village."

Coertze said, "No lights can be seen from the village."

"All the same, we must have a cover," I said. "If we have to hang around in the vicinity for at least one day then we must have a sound reason. Has anyone got any ideas?"

There was a silence and suddenly Walker spoke for the first time. "What about a car and a caravan? The English are noted for that kind of thing—camping and so on. The Italians don't even have a word for it, they use the English word. If we camp out for a couple of nights we'll be only another English crowd as far as the peasants are concerned."

We all thought about that and it seemed a good idea. The Contessa said, "I can arrange for the car and the caravan and a tent."

I started to tick off all the things we would need. "We want lights."

"We use the headlights of the car," said Coertze.

"That's for outside," I said. "We'll need lights for inside. We'll need torches—say a dozen—and lots of torch cells." I nodded to Morese. "You get those. We need picks and shovels, say four of each. We'll need lorries. How many to do the job in one haul?"

"Two three-tonners," said Coertze with certainty. "The Germans had four, but they were carrying a lot of stuff we won't want."

"We'll have to have those standing by with the drivers," I said. "Then we'll need a lot of timber to make crates. The gold will need re-boxing."

"Why do that when it's already in boxes?" objected Coertze. "It's just a lot of extra work."

"Think back," I said patiently. "Think back to the first time you saw those boxes in the German truck. You *recognised* them as bullion boxes. We don't want any snooper doing the same on the way back."

Walker said, "You don't have to take the gold out, and it wouldn't need much timber. Just nail thin pieces of wood on the outside of the bullion boxes to change their shape and make them look different."

Walker was a real idea machine when he wasn't on the drink.

He said, "There must be plenty of timber down there we can use."

"No," I said, "We use new wood. I don't want anything that looks or even smells as though it's come from a hole in the ground. Besides, there might be a mark on the wood we could miss which would give the game away."

"You don't take any chances, do you?" observed the Contessa.

"I'm not a gambler," I said shortly. "The timber can go up in the trucks." I looked at Morese.

"I will get it," he said.

"Don't forget hammers and nails," I said. I was trying to think of everything. If we slipped up on this job it would be because of some insignificant item which nobody had thought important.

There was a low, repeated whistle from the dockside. Morese looked at the Contessa and she nodded almost imperceptibly. He got up and went on deck.

I said to Coertze, "Is there anything else we ought to know—anything you've forgotten or left out?"

"No," he said. "That's all."

Morese came back and said to the Contessa, "He wants to talk to you."

She rose and left the cabin and Morese followed her on deck. Through the open port I could hear a low-voiced conversation.

"I don't trust them," said Coertze violently. "I don't trust that bitch and I don't trust Morese. He's a bad bastard; he was a bad bastard in the war. He didn't take any prisoners —according to him they were all shot while escaping."

"So were yours," I said, "when you took the gold."

He bridled. "That was different; they *were* escaping."

"Very conveniently," I said acidly. It galled me that this man, whom I had good reason to suspect of murdering at least four others, should be so mealy-mouthed.

He brooded a little, then said, "What's to stop them taking it all from us when we've got it out? What's to stop them shooting us and leaving us in the tunnel when they seal it up again?"

"Nothing that you'd understand," I said. "Just the feeling of a girl for her father and her family." I didn't elaborate on that; I wasn't certain myself that it was a valid argument.

The Contessa and Morese came back. She said, "Two of Torloni's men are in Rapallo. They were asking the Port Captain about you not ten minutes ago."

I said, "Don't tell me that the Port Captain is one of your friends."

"No, but the Chief Customs Officer is. He recognised them immediately. One of them he had put in jail three years ago for smuggling heroin; the other he has been trying to catch for a long time. Both of them work for Torloni, he says."

"Well, we couldn't hope to hide from them indefinitely," I said. "But they mustn't connect you with us—not yet, anyway—so you'll have to wait until it's dark before you leave."

She said, "I am having them watched."

"That's fine, but it's not enough," I said. "I want to do to Metcalfe what he's been doing to us. I want Torloni watched in Genoa; I want the docks watched all along this coast for Metcalfe's boat. I want to know when he comes to Italy." I gave her a detailed description of Metcalfe, of Krupke and the Fairmile. "Can you do all that?"

"Of course. You will know all about this Metcalfe as soon as he sets foot in Italy."

"Good," I said. "Then what about a drink?" I looked at Coertze. "It seems you didn't scare Metcalfe off, after all." He looked back at me with an expressionless face, and I laughed. "Don't look so glum. Get out the bottle and cheer up."

V

We didn't see the Contessa or Morese after that. They stayed out of sight, but next morning I found a note in the cockpit telling me to go to the Three Fishes and say that I wanted a watchman for *Sanford*.

I went, of course, and Giuseppi was more friendly than when I had last seen him. He served me personally and, as he put down the plate, I said, "You ought to know what goes on on the waterfront. Can you recommend a watchman for my boat? He must be honest."

"Ah, yes, signor," he said. "I have the very man—old Luigi there. It is a pity; he was wounded during the war and since then he has been able to undertake only light work. At present he is unemployed."

"Send him over when I have finished breakfast," I said.

Thus is was that we got an honest watchman and old Luigi became the go-between between the Contessa and *Sanford*.

Every morning he would bring a letter in which the Contessa detailed her progress.

Torloni was being watched, but nothing seemed to be happening; his men were still in Rapallo watching *Sanford* and being watched themselves; the trucks had been arranged for and the drivers were ready; the timber was prepared and the tools had been bought; she had been offered a German caravan but she had heard of an English caravan for sale in Milan and thought it would be better—would I give her some money to buy it as she had none.

It all seemed to be working out satisfactorily.

The three of us from *Sanford* spent our time sight-seeing, much to the disgust of Torloni's spies. I spent a lot of time in the Yacht Club and it was soon noised about that I intended to settle in the Mediterranean and was looking for a suitable boatyard to buy.

On our fifth day in Rapallo the morning letter instructed me to go to the boatyard of Silvio Palmerini and to ask for a quotation for the slipping and painting of *Sanford*. "The price will be right," wrote the Contessa. "Silvio is one of my—our—friends."

Palmerini's yard was some way out of Rapallo. Palmerini was a gnarled man of about sixty who ruled his yard and his three sons with soft words and a will of iron. I said, "You understand, Signor Palmerini, that I am a boat-builder, too. I would like to do the job myself in your yard."

He nodded. It was only natural that a man must look after his own boat if he could; besides, it would be cheaper.

"And I would want it under cover," I said. "I fastened the keel in an experimental way and I may want to take it off to see if it is satisfactory."

He nodded again. Experimental ways were risky and a man should stick to the old traditional ways of doing things. It would be foolish, indeed, if milord's keel dropped off in the middle of the Mediterranean.

I agreed that I should look a fool, and said, "My friends and I are capable of doing the work and we shall not need extra labour. All that is required is a place where we can work undisturbed."

He nodded a third time. He had a large shed we could use and which could be locked. No one would disturb us, not even himself—certainly no one outside his family—he would see to that. And was milord the rich Englishman who wanted to buy a boatyard? If so, then perhaps the milord would consider the boatyard Palmerini, the paragon of the Western Mediterranean.

That brought me up with a jerk. Another piece of polite blackmail was under way and I could see that I would have to buy the yard, probably at an exorbitant price—the price of silence.

I said diplomatically, "Yes, I am thinking of buying a yard, but the wise man explores every avenue." Dammit, I was falling into his way of speech. "I have been to Spain and France; now I am in Italy and after Italy I am going to Greece. I must look at everything."

He nodded vigorously, his crab-apple head bobbing up and down. Yes, the milord was indeed wise to look at everything, but in spite of that he was sure that the milord would unfailingly return to the boatyard Palmerini because it was certainly the best in the whole Mediterranean.

Pah, what did the Greeks know of fine building? All they knew were their clumsy caiques. The price would be reasonable for milord since it appeared that they had mutual friends, and such price could be spread over a period provided the proper guarantees could be given.

From this I understood the old rascal to say that he would wait until the whole job was completed and I had fluid capital, if I could prove that I would keep my word.

I went back to *Sanford* feeling satisfied that this part of the programme was going well. Even if I had to buy Palmerini's yard, it would not be a bad thing and any lengthening of the price could be written off as expedition expenses.

On the ninth day of our stay in Rapallo the usual morning letter announced that all was now ready and we could start at any time. However, it was felt that, since the next day was Sunday, it would be more fitting to begin the expedition inland on Monday. That gave an elevating tone to the whole thing, I thought; another crazy aspect of a crazy adventure.

The Contessa wrote: "Torloni's men will be discreetly taken care of, and will not connect their inability to find you with any trickery on your part. They will have no suspicions. Leave your boat in the care of Luigi and meet me at nine in the morning at the Three Fishes."

I put a match to the letter and called Luigi below. "They say you are an honest man, Luigi; would you take a bribe?"

He was porperly horrified. "Oh no, signor."

"You know this boat is being watched?"

"Yes, signor. They are enemies of you and Madame."

"Do you know what Madame and I are doing?"

He shook his head. "No, signor. I came because Madame said you needed my help. I did not ask any questions," he said with dignity.

I tapped on the table. "My friends and I are going away for a few days soon, leaving the boat in your charge. What will you do if the men who are watching want to bribe you to let them search the boat?"

He drew himself up. "I would slap the money out of their hands, signor."

"No, you won't," I said. "You will say it is not enough and you will ask for more money. When you get it, you will let them search the boat."

He looked at me uncomprehendingly. I said slowly, "I don't mind if they search—there is nothing to be found. There is no reason why you should not make some money out of Madame's enemies."

He laughed suddenly and slapped his thigh. "That is good, signor; that is very good. You *want* them to search."

"Yes," I replied. "But don't make it too easy for them or they will be suspicious."

I wanted, as a last resort, to try to fool Metcalfe as I had fooled him in Barcelona, or rather, as I had hoped to fool him before Coertze put his foot in it. I wrote a letter to the Contessa telling her what I was doing, and gave it to Luigi to pass on.

"How long have you known Madame?" I asked curiously.

"Since the war, signor, when she was a little girl."

"You would do anything for her, wouldn't you?"

"Why not?" he asked in surprise. "She has done more for me that I can ever repay. She paid for the doctors after the war when they straightened my leg. It is not her fault they could not get it properly straight—but I would have been a cripple, otherwise."

This was a new light on Francesca. "Thank you, Luigi," I said. "Give the letter to Madame when you see her."

I told Coertze and Walker what was happening. There was nothing else to do now but wait for Monday morning.

V. THE TUNNEL

On Monday morning I again set the stage, leaving papers where they could easily be found. On the principle of the Purloined Letter I had even worked out a costing for a refit of *Sanford* at Palmerini's boatyard, together with some estimates of the probable cost of buying the yard. If we were seen there later we would have good reason.

We left just before nine, saying good-bye to Luigi, who gave me a broad wink, and arrived at the Three Fishes on time. The Contessa and Morese were waiting and we joined them for breakfast. The Contessa wore clothing of an indefinably English cut of which I approved; she was using her brains.

I said, "How did you get rid of Torloni's boys?"

Morese grinned. "One of them had an accident with his car. The other, who was waiting for him at the dock, got tired of waiting and unaccountably fell into the water. He had to get a taxi to his hotel so that he could change his clothes."

"You're friend Metcalfe arrived in Genoa last night," said the Contessa.

"You're sure."

"I'm certain. He went straight to Torloni and stayed with him for a long time. Then he went to a hotel."

That settled that. I had wondered for a long time if my suspicions of Metcalfe hadn't been just a fevered bit of imagination. After all, my whole case against Metcalfe had been built up of supposition and what I knew of his character.

"You're having him watched?"

"Of course."

Breakfast arrived and all conversation stopped until Giuseppi went back to his counter. Then I said, "All right, friend Kobus, this is where you tell us where the gold is."

Coertze's head came up with a jerk. "Not on," he said. "I'll take you there, but I'm not telling first."

I sighed. "Look, these good people have laid on transport. How can they tell the trucks where to rendezvous unless we know where we're going?"

"They can telephone back here."

"From where?"

"There'll be a phone in the village."

"None of us is going anywhere near that village," I said. "Least of all one of us foreigners. And if you think I'll let one of these two go in alone, you're crazy. From now on we don't let either of them out of our sight."

"Not very trusting, are you?" observed the Contessa.

I looked at her. "Do you trust me?"

"Not much."

"Then we're even." I turned back to Coertze. "Any telephoning the Contessa is going to do is from that telephone in the corner there—with me at her elbow."

"Don't call me the Contessa," she snapped.

I ignored her and concentrated on Coertze. "So, you see, we have to know the spot. If you won't tell us, I'm sure that Walker will—but I'd rather it was you."

He thought about it for quite a while, then he said, "*Magtig*, but you'll argue your way into heaven one day. All right, it's about forty miles north of here, between Varsi and Tassaro." He went into detailed explanations and Morese said, "It's right in the hills."

I said, "Do you think you can direct the trucks to this place?"

Francesca said, "I will tell them to wait in Varsi. We will not need them until the second night; we can go to Varsi and direct them from there to-morrow."

"O.K.," I said. "Let's make that phone call."

I escorted her to the corner and stood by while she gave the instructions, making sure she slipped nothing over. A trustful lot, we were. When we got back to the table, I said, "That does it; we can start at any time."

We finished breakfast and got up to go. Francesca said, "Not by the front; Torloni's men will be back now and they can see this café. We go this way."

She led us out by the back door into a yard where a car was standing with an Eccles touring caravan already coupled. She said, "I stocked up with enough food for a week—it might be necessary."

"It won't," I said grimly. "If we don't have to stuff out by to-morrow night we'll never get it—not with Metcalfe sniffing on our trail."

I looked at our party and made a quick decision. "We look English enough, all except you, Morese; you just don't fit. You travel in the caravan and keep out of sight."

He frowned and looked at Francesca. She said, "Get into

the caravan, Piero; do as Mr. Halloran says," and then turned to me. "Piero takes his instructions from no onc but me, Mr. Halloran. I hope you remember that in future."

I shrugged and said, "Let's go."

Coertze was driving because he knew the way. Walker was also in front and Francesca and I shared the back seat. No one did much talking and Coertze drove very slowly because he was unaccustomed to towing a caravan and driving on the right simultaneously.

We left Rapallo and were soon ascending into the hills—the Ligurian Appennines. It looked poor country with stony soil and not much cultivation. What agriculture there was was scattered and devoted to vines and olives, the two trees which look as though they've been tortured to death. Within the hour we were in Varsi, and soon after that, we left the main road and bounced along a secondary country road, unmetalled and with a poor surface. It had not rained for some days and the dust rose in clouds.

After a while Coertze slowed down almost to a stop as he came to a corner. "This is where we shot up the trucks," he said.

We turned the corner and saw a long stretch of empty road. Coertze stopped the car and Walker got out. This was the first time he had seen the place in fifteen years. He walked a little way up the road to a large rock on the right, then turned and looked back. I guessed it was by that rock that he had stood while he poured bullets into the driver of the staff car.

I thought about the sudden and dreadful slaughter that had happened on that spot and, looking up the shaggy hillside, I visualised the running prisoners being hunted and shot down. I said abruptly, "No point in waiting here, let's get on with it."

Coertze put the car into gear and drove forward slowly until Walker had jumped in, then he picked up speed and we were on our way again. "Not far now," said Walker. His voice was husky with excitement.

Less than fifteen minutes later Coertze pulled up again at the junction of another road so unused that it was almost invisible. "The old mine is about a mile and a half up there," he said. "What do we do now?"

Francesca and I got out of the car and stretched our stiffened legs. I looked about and saw a stream about a hundred yards away. "That's convenient," I said. "The perfect camp site. One thing is certain—none of us so much as looks sideways at that side road during the hours of daylight."

We pulled the caravan off the road and extended the balance legs, then we put up the tent. Francesca went into the caravan and talked to Morese. I said, "Now, for God's sake, let's act like innocent tourists. We're mad Englishmen who prefer to live uncomfortably rather than stay at a hotel."

It was a long day. After lunch, which Francesca made in the little galley of the caravan, we sat about and talked desultorily and waited for the sun to go down. Francesca stayed in the caravan most of the time keeping Morese company; Walker fidgeted; Coertze was apparently lost in contemplating his navel; I tried to sleep, but couldn't.

The only excitement during the afternoon was the slow approach of a farm cart. It hove into sight as a puff of dust at the end of the road and gradually, with snail-like pace, came near enough to be identified. Coertze roused himself enough to make a number of small wagers as to the time it would draw level with the camp. At last it creaked past, drawn by two oxen and looking like a refugee from a Breughel painting. A peasant trudged alongside and I mustered my worst Italian, waved and said, "*Buon giorno.*"

He gave me a sideways look, muttered something I did not catch, and went on his way. That was the only traffic on the road the whole time we were there.

At half past four I roused myself and went to the caravan to see Francesca. "We'd better eat early," I said. "As soon as it's dark we'll be taking the car to the mine."

"Everything is in cans," she said. "It will be easy to prepare. We will want something to eat during the night, so I got two of these big vacuum containers—I will cook the food before we go and it will keep hot all night. There are also some vacuum flasks for coffee."

"You've been spending my money well," I said.

She ignored that. "I will need some water. Will you get some from the stream?"

"If you will come with me," I said. "You need to stretch a bit." I had a sudden urge to talk to her, to find out what made her tick.

"All right," she said, and opening a cupboard, produced three canvas buckets. As we walked towards the stream, I said, "You must have been very young during the war."

"I was. We took to the hills, my father and I, when I was ten years old." She waved at the surrounding mountains. "These hills."

"Not a very pleasant life for a little girl."

She considered that. "It was fun at first. Everyone likes

a camping holiday and this was one long holiday for me. Yes, it was fun."

"When did it stop being fun?"

Her face was quietly sad. "When the men started to die; when the fighting began. Then it was not fun, it became a serious thing we were doing. It was a good thing—but it was terrible."

"And you worked in the hospital?"

"Yes. I tended Walker when he came from the prison camp. Did you know that?"

I remembered Walker's description of the grave little girl who wanted him to get better so he could kill Germans. "He told me," I said.

We reached the stream and I looked at it doubtfully. It looked clear enough, but I said, "Is it all right for drinking?"

"I will boil the water; it will be all right," she said, and knelt to dig a hole in the shallows. "We must have a hole deep enough to take a bucket; it is easier then."

I helped her make a hole, reflecting that this was a product of her guerilla training. I would have tried to fill the buckets in drips and drabs. When the hole was big enough we sat on the bank waiting for the sediment to settle, and I said, "Was Coertze ever wounded?"

"No, he was very lucky. He was never wounded beyond a scratch, although there were many times he could have been."

I offered her a cigarette and lit it. "So he did a lot of fighting?"

"All the men fought," she said, and drew on the cigarette reflectively. "But Coertze seemed to *like* fighting. He killed a lot of Germans—and Italians."

"What Italians?" I said quickly. I was thinking of Walker's story.

"The Fascists," she said. "Those who stuck by Mussolini during the time of the Salo Republic. There was a civil war going on in these mountains. Did you know that?"

"No, I didn't," I said. "There's a lot about Italy that I don't know."

We sat quietly for a while, then I said, "So Coertze was a killer?"

"He was a good soldier—the kind of man we needed. He was a leader."

I switched. "How was Alberto killed?"

"He fell off a cliff when the Germans were chasing Umberto's section. I heard that Coertze nearly rescued him, but didn't get there in time."

" Um," I said. " I heard it was something like that. How did Harrison and Parker die?"

She wrinkled her brow. " Harrison and Parker? Oh yes, they were in what we called the Foreign Legion. They were killed in action. Not at the same time, at different times."

" And Donato Rinaldi; how was he killed?"

" That was a funny thing. He was found dead near the camp with his head crushed. He was lying under a cliff and it was thought that he had been climbing and had fallen off."

" Why should he climb? Was he a mountaineer or something like that?"

" I don't think so, but he was a young man and young men do foolish things like that."

I smiled, thinking to myself; not only the very young are foolish; and tossed a pebble into the stream. " It sounds very like the song about the ' Ten Little Niggers.' ' And then there were Two.' Why did Walker leave?"

She looked up sharply. " Are you saying that these men should not have died? That someone from the camp killed them?"

I shrugged. " I'm not saying anything—but it was very convenient for someone. You see, six men hid this gold and four of them came to a sudden end shortly afterwards." I tossed another pebble into the water. " Who profits? There are only two—Walker and Coertze. Why did Walker leave?"

" I don't know. He left suddenly. I remember he told my father that he was going to try to join the Allied armies. They were quite close at that time."

" Was Coertze in the camp when Walker left?"

She thought for a long time, then said, " I don't know; I can't remember."

" Walker says he left because he was frightened of Coertze. He still is, for that matter. Our Kobus is a very frightening man, sometimes."

Francesca said slowly, " There was Alberto on the cliff. Coertze could have . . ."

" . . . pushed him off? Yes, he could. And Walker said that Parker was shot in the *back* of the head. By all accounts, including your, Coertze is a natural-born killer. It all adds up."

She said, " I always knew that Coertze was a violent man, but . . ."

" But? Why don't you like him, Francesca?"

She threw the stub of her cigarette into the water and watched it float downstream. " It was just one of those things that happen between a man and a woman. He was . . . too pressing."

"When was this?"

"Three years ago. Just after I was married."

I hesitated. I wanted to ask her about that marriage, but she suddenly stood up and said, "We must get the water."

As we were going back to the caravan I said, "It looks as though I'll have to be ready to jump Coertze—he could be dangerous. You'd better tell Piero the story so that he can be prepared if anything happens."

She stopped. "I thought Coertze was your friend. I thought you were on his side."

"I'm on nobody's side," I said shortly. "And I don't condone murder."

We walked the rest of the way in silence.

For the rest of the afternoon until it became dark Francesca was busy cooking in the caravan. As the light faded the rest of us began to make our preparations. We put the picks and shovels in the boot of the car, together with some torches. Piero had provided a Tilley pressure lamp together with half a gallon of paraffin—that would be a lot better than torches once we got into the tunnel. He also hauled a wheelbarrow out of the caravan, and said, "I thought we could use this for taking the rock away; we must not leave loose rock at the entrance of the tunnel."

I was pleased about that; it was something I had forgotten.

Coertze examined the picks with a professional air, but found no fault. To me, a pick is a pick and a spade is a bloody shovel, but I suppose that even pick-and-shovelling has its more erudite technicalities. As I was helping Piero put the wheelbarrow into the boot my foot turned on a stone and I was thrown heavily against Coertze.

"Sorry," I said.

"Don't be sorry, be more careful," he grunted.

We got the wheelbarrow settled—although the top of the boot wouldn't close—and I said to Coertze in a low voice, "I'd like to talk to you . . . over there."

We wandered a short distance from the rest of the party where we were hidden in the gathering darkness. "What is it?" asked Coertze.

I tapped the hard bulge under the breast of his jacket, and said, "I think that's a gun."

"It is a gun," he said.

"Who are you thinking of shooting?"

"Anyone who gets between me and the gold."

"Now listen carefully," I said in a hard voice. "You're not going to shoot anyone, because you're going to give that gun to me. If you don't, you can get the gold yourself. I didn't come to Italy to kill anybody; *I'm* not a murderer."

Coertze said, "*Klein man,* if you want this gun you'll have to take it from me."

"O.K. You can force us all up to the mine at pistol point. But it's dark and you'll get a rock thrown at your head as soon as you turn your back—and I'd just as soon be the one who throws it. And if you get the gold out—at pistol point—what are you going to do besides sit on it? You can't get it to the coast without Francesca's men and you can't get it out of Italy without me."

I had him cornered in the same old stalemate that had been griping him since we left South Africa. He was foxed and he knew it.

He said, "How do we know the Contessa's partisans aren't hiding in these damned hills waiting to jump us as soon as the tunnel is opened?"

"Because they don't know where we are," I said. "The only instruction that the truck drivers had was to go to Varsi. Anyway, they wouldn't try to jump us; we have the Contessa as hostage."

He hesitated, and I said, "Now, give me the gun."

Slowly he put his hand inside his jacket and pulled out the gun. It was too dark to see his eyes but I knew they were filled with hate. He held the gun pointed at me and I am sure he was tempted to shoot—but he relaxed and put it into my outstretched hand.

"There'll be a big reckoning between us when this is all over," he said.

I remained silent and looked at the gun. It was a Luger, just like my own pistol which I had left in South Africa. I held it on him, and said, "Now stand very still; I'm going to search you."

He cursed me, but stood quietly while I tapped his pockets. Sure enough, in his jacked pocket I found a spare magazine. I took the clip from the Luger and snapped the action to see if he had a round up the spout. He had.

He said, "Morese is sure to have a gun."

"We'll see about that right now," I said. "I'll tackle him and you stand behind him ready to sock him."

We walked back to the caravan and I called for Francesca

and Piero and when they came Coertze unobtrusively stationed himself behind the big Italian. I said to Francesca, "Has Piero got a gun?"

She looked startled. "I don't know." She turned to him. "Are you carrying a gun, Piero?"

He hesitated, then nodded. I brought up the Luger and held it on him. "All right, bring it out—slowly."

He looked at the Luger and his brows drew down angrily, but he obeyed orders and slowly pulled a gun from a shoulder holster. I said, "This is one time you take orders from me, Piero. Give it to Francesca."

He passed the pistol to Francesca and I put the Luger away and took it from her. It was an army Beretta, probably a relic of his partisan days. I took the clip out, worked the action and put it in another pocket. Coertze passed two spare clips to me which he had taken from Piero's pockets.

I said to Walker, "Are you carrying a gun?"

He shook his head.

"Come and be searched." I was taking no chances.

Walker was bare of guns, so I said, "Now search the car and see if anything is tucked away there."

I turned to Francesca. "Are you carrying anything lethal?"

She folded her arms. "Are you going to search me, too?"

"No. I'll take your word, if you'll give it."

She dropped her aggressive pose. "I haven't a gun," she said in a low voice.

I said, "Now listen, everybody. I've taken a gun from Coertze and a gun from Morese. I hold in my hands the ammunition for those guns." With a quick double jerk I threw the clips away into the darkness and they clattered on a rock. "If there's going to be any fighting between us it will be with bare fists. Nobody gets killed, do you hear?"

I took the empty pistols from my pockets and gave them back to Coertze and Piero. "You can use these as hammers to nail the crates up."

They took them with bad grace and I said, "We've wasted enough time with this nonsense. Is that car ready?"

"Nothing in here," said Walker.

As the others were getting into the car, Francesca said to me, "I'm glad you did that. I didn't know Piero had a gun."

"I didn't know Coertze had one, either; although I should have guessed—knowing his record."

"How did you take it from him?" she asked curiously.

"Psychology," I said. "He would rather have the gold than

kill me. Once he gets the gold is might be a different matter."

"You will have to be very careful," she said.

"It's nice to know you care," I said. "Let's get in the car."

II

Coertze drove slowly without lights along the overgrown road until we had turned a corner and were out of sight of the "main" highway. I could hear the long grass swishing on the underside of the car. Once the first corner was turned he switched on the lights and picked up speed.

No one spoke. Coertze and Morese were mad at me and so was Francesca because of what I'd said. Walker was boiling with ill-suppressed excitement, but he caught the mood of the others and remained quiet. I said nothing because I had nothing to say.

It didn't take long to get to the mine and soon the headlights swept over the ruins of buildings—the shabby remnants of an industrial enterprise. There is nothing more ruinous-looking than derelict factory buildings and neglected machinery. Not that there was much left. The surrounding peasantry must have overrun the place like a swarm of locusts very soon after the mine was abandoned and carried off everything of value. What was left was worth about ten lire and would have cost a hundred thousand lire to take away.

Coertze stopped the car and we all got out. Piero said, "What kind of a mine was this?"

"A lead mine," said Coertze. "It was abandoned a long time go—about 1908, I was told."

"That was about the time they found the big deposits in Sardinia," said Piero. "It was easier to ship ore to the smeltery in Spezia than to rail it from here."

"Where's your tunnel?" I asked.

Coertze pointed. "Over there. There were four others besides the one I blocked."

"We might as well get the car into position," I said, so Coertze got into the driving seat and edged the car forward. The beams of light swept round and illuminated the caved-in mouth of the tunnel. It looked as though it would need a regiment of pioneers to dig that lot away and it would probably take them a month.

Coertze leaned out of the side window. "I did a good job there," he said with satisfaction.

I said, "You're sure we can get through there in one night?"

"Easy," he said.

I suppose he knew what he was about—he had been a miner. I went to help Piero and Walker get the tools from the boot and Coertze went to the rockfall and began to examine it. From this time on he took charge and I let him—I knew nothing about the job and he did. His commands were firm-voiced and we all jumped to it with a will.

He said, "We don't have to dig the whole lot away. I set the charges so that the fall on this side would be fairly thin—not more than ten feet."

I said, "Ten feet sounds like a hell of a lot."

"It's nothing," he said, contemptuous of my ignorance. "It isn't as though it was solid rock—this stuff is pretty loose." He turned and pointed. "Behind that building you'll find some baulks of timber I sorted out three years ago. You and Morese go and get them. Walker and I will start to dig this stuff out."

"What can I do?" asked Francesca.

"You can load up the wheelbarrow with the stuff we dig out. Then take it away and scatter it so that it looks natural. Morese is right—we don't want to leave a pile of rocks here."

Piero and I took torches and found the timber where Coertze had indicated. I thought of Coertze coming here every three or four years, frustrated by a problem he couldn't solve. He must have planned this excavating problem many times and spent hours sorting out this timber in readiness for a job which might never have happened. No wonder he was so touchy.

It took us about an hour to transfer all the timber and by that time Coertze and Walker had penetrated three feet into the rockfall. That was good going, and I said as much. Coertze said, "It won't be as easy as this all the way. We'll have to stop and shore the roof; that'll take time."

The hole he was digging was not very big; about five feet high and two feet wide—just room enough for one man to go through. Coertze began to select his timbers for the shoring and Piero and I helped Francesca to distribute the spoil.

Coertze was right. The shoring of the roof took a long time but it had to be done. It would be bad if the whole thing collapsed and we had to begin all over again; besides, someone might get hurt. A moon rose, making the distribution of the spoil easier, so the car lights were switched off and Coertze was working by the light of the Tilley lamp.

He would not let anyone work at the face except himself,

so Walker, Piero and I took it in turns helping him, standing behind him and passing out the loose rocks to the entrance of the passage. After another three hours we had six feet of firmly shored passage drilled through the rockfall and at this stage we broke off for something to eat.

Piero had spoken to me about taking away his gun. He said, "I was angry when you did it. I do not like to have guns pointed at me."

"It was empty," I said.

"That I found out, and it was that which made me angrier." He chuckled suddenly. "But I think it was well done, now I have thought about it. It is best if there is no shooting."

We were some distance from the rockfall. I said, "Did Francesca tell you about Coertze?"

"Yes. She told me what you said. It is something I have not thought of at all. I was surprised when Donato Rinaldi was found dead that time during the war, but I did not think anyone would have killed him. We were all friends."

Gold is a solvent which dissolves friendships, I thought, but I could not put that into my limited Italian. Instead, I said, "From what you know of that time, do you think that Coertze could have killed these four men?"

Piero said, "He could not have killed Harrison because I myself saw Harrison killed. He was shot by a German and I killed the man who shot him. But the others—Parker, Corso and Rinaldi—yes, I think Coertze could have killed them. He was a man who thought nothing of killing."

"He could have killed them, but did he?" I asked.

Piero shrugged. "Who can tell? It was a long time ago and there are no witnesses."

That was that, and there seemed no point in pressing it, so we returned to our work.

Coertze hurried over his meal so that he could get back to the rock face. His eyes gleamed brightly in the light of the lamp; the lust for gold was strong upon him, for he was within four feet of the treasure for which he had been waiting fifteen years. Walker was as bad; he scrambled to his feet as soon as Coertze made a move and they both hurried to the rockfall.

Piero and Francesca were more placid. They had not seen the gold and mere descriptive words have not that immediacy. Francesca leisurely finished her midnight snack and then collected the dishes and took them to the car.

I said to Piero, "That is a very strange woman."

"Any child who was brought up in a guerilla camp would be different," he said. "She has had a difficult life."

I said carefully, " I understand she has had an unfortunate marriage."

He spat. " Estrenoli is a degenerate."

" Then why did she marry him?"

" The ways of the *aristos* are not our ways," he said. " It was an arranged marriage—or so everyone thinks. But that was not really the way of it."

" What do you mean?"

He accepted a cigarette. " Do you know what the Communists did to her father?"

" She told me something about it."

" It was shameful. He was a man, a true man, and they were not fit to lick his boots. And now he is but a shell, an old broken man." He struck a match and the flame lit up his face. " Injustice can crush the life from a man even if his body still walks the streets," he said.

" What has this to do with Francesca's marriage?"

" The old man was against it. He knew the Estrenoli breed. But Madame was insistent on it. You see, young Estrenoli wanted her. There was no love in him, only lust—Madame is a very beautiful woman—and so he wanted her, but he could not get her. She knew what he was."

This was confusing. " Then why the hell did she marry him?"

" That was where Estrenoli was clever. He has an uncle in the Government and he said that perhaps they would reconsider the case of her father. But, of course, there was a price."

" I see," I said thoughtfully.

" So she married him. I would as soon she married an animal."

" And he found he could not keep his promise?"

" Could not?" said Piero disgustedly. " He had no intention of keeping it. The Estrenolis have not kept a promise in the last five hundred years." He sighed. " You see, she is a good daughter of the Church and when she married him, Estrenoli knew that he had her for ever. And he was proud of her; oh yes, very proud. She was the most beautiful woman in Roma, and he bought her clothes and dressed her as a child will dress a doll. She was the most expensively dressed mannequin in Italy."

" And then?"

" And then he got tired of her. He is an unnatural man and he went back to his little boys and his drugs and all the other vices of Roma. Signor Halloran, Roman society is the most corrupt in the world."

I had heard something of that; there had been a recent case of a drowned girl which threatened to rip apart the whole shoddy mess. But it was said that the Italian Government was intent on hushing it up.

Piero said, " At that time she helped her father and her old comrades. There were many cases of hardship and she did what she could. But Estrenoli found out and said he would not have his money squandered on a lot of filthy partisans, so he did not give her any more money—not one single lire. He tried to corrupt her, to bring her down to his level, but he could not—she is incorruptible. So then he threw her out on to the street—he had what he wanted, as much as he could get, and he was finished with her."

" So she came back to Liguria."

" Yes. We help her when we can because of what she is and because of her father. We also try to help him, but that is difficult because he refuses to accept what he calls charity."

" And she is still married to Estrenoli?"

" There is no divorce in Italy and she follows the Church. But before God I say the Church is wrong when this can happen."

I said, " And so you are helping her in this venture."

" I think it is wrong and I think she is mistaken," he said. " I think many lives will be lost because of this. But I am helping her."

" This is what is puzzling me," I said. " Her father is an old man; this gold cannot help him much."

" But it is not only for her father," said Piero. " She says that the money is for all the men who fought with her father and were cheated by the Communists. She says it will be used to send them to hospitals when that is necessary and to educate their children. It will be a good thing if there is no killing."

" Yes, it will," I said reflectively. " I do not want killing, either, Piero."

" I know, Signor Halloran; you have already shown that. But there are others—Torloni and this Metcalfe. And there is your friend Coertze."

" You don't trust him, do you? What about Walker?"

" Pah—a nonentity."

" And me? Do you trust me?"

He stood and put a foot on his cigarette deliberately. " I would trust you in another place, Signor Halloran, such as in a boat or on a mountain. But gold is not good for the character."

He had said in different words what I had thought earlier.

I was going to reply when Coertze shouted irately, "What the hell are you doing out there? Come and get this stuff away."

So we went on with the work.

III

We broke through at three in the morning. Coertze gave a joyous shout as his pick point disappeared unresistingly into emptiness. Within ten minutes he had broken a hole big enough to crawl through and he went into it like a terrier after a rabbit. I pushed the Tilley lamp through the hole and followed it.

I found Coertze scrambling over fallen rocks which littered the floor of the tunnel. "Hold on," I said. "There's no hurry."

He took no notice but plunged on into the darkness. There was a clang and he started to swear. "Bring that bloody light," he shouted.

I moved forward and the circle of light moved with me. Coertze had run full tilt into the front of a truck. He had gashed his cheek and running blood was making runnels in the dust which coated his face, giving him a maniacal look which was accentuated by the glare of his eyes.

"Here it is," he cackled. "*Magtig,* what did I tell you? I told you I had gold here. Well, here it is, as much gold as comes out of the Reef in a month." He looked at me in sudden wonder. "*Christus,* but I'm happy," he said. "I never thought I'd make it."

I could hear the others coming through the hole and I waited for them to come up. "Kobus Coertze is going to give us a guided tour of his treasure cave," I said.

Walker said, chattering, "The gold is in the first truck, this one. Most of it, that is. There's some more, though, in the second one, but most of it is in this one. The jewels are in the second one; lots and lots of necklaces and rings, diamonds and emeralds and pearls and cigarette lighters and cases, all in gold, and there's lots of money, too, lire and dollars and pounds and stuff like that, and there's lots of papers but those are in the trucks right at the back with the bodies . . ." His voice trailed off. "With the bodies," he repeated vacantly.

There was a bit of a silence then as we realised that this was a mausoleum as well as a treasure cave. Coertze recovered his usual gravity and took the lamp from me. He held it up and looked at the first truck. "I should have put it up on blocks," he observed wryly.

The tyres were rotten and sagging, as flat as I've ever seen tyres. "You know," said Coertze. "When we put this lot in here, my intention was to drive these trucks out some time. I never thought it would be fifteen years." He gave a short laugh. "We'd have a job starting these engines now."

Walker said impatiently. "Well, let's get on with it." He had apparently recovered from the scare he had given himself.

I said, "We'd better do this methodically, truck by truck. Let's have a look in the first one."

Coertze led the way, holding up the Tilley lamp. There was just enough room to squeeze between the truck and the side of the tunnel. I noticed the shattered windscreen where a burst of machine-gun fire had killed the driver and his mate. Everything was covered with a heavy layer of dust, most of which must have been deposited when Coertze originally blew in the front of the tunnel.

Coertze was hammering at the bolts of the tailboard with a piece of rock. "The damn' things have seized solid," he said. "I'll need a hammer."

"Piero," I called. "Bring a hammer."

"I've got one," said Francesca quietly, so close behind me that I jumped. I took it and passed it on to Coertze. With a few blows the bolt came free and he attacked the other and caught the tailboard as it dropped. "Right," he said, "here we go for the gold," and vaulted into the truck.

I handed him the lamp and then climbed up and turned to give Francesca a hand. Walker crowded past me, eager to see the gold, while Piero climbed in more sedately. We squatted on our haunches in a circle, sitting on the bullion boxes.

"Where's the one we opened?" asked Coertze. "It must be at the back somewhere."

Francesca gave a yelp. "I've got a nail in my foot."

"That's the box," said Coertze with satisfaction.

Francesca moved and Coertze held up the lamp. The box on which Francesca had been sitting had been torn open and the cover roughly replaced. I stretched my hand and lifted the lid slowly. In the light of the lamp there was the yellow gleam of metal, the dull radiance of gold which rusts not nor doth moth corrupt—rather like treasure laid up in heaven. This gold, however, had been laid up in hell.

Coertze sighed. "There it is."

I said to Francesca, "Did you hurt your foot?"

She was staring at the gold. "No, it's all right," she said absently.

Piero lifted an ingot from the box. He misjudged the weight and tried to use one hand; then he got both hands to it and rested the ingot on his thighs. "It *is* gold!" he said in wonder.

The ingot was passed round the circle and we all handled it and stroked it. I felt a sudden resurgence of the passion I had felt in Aristide's strong-room when I held the heavy gold Hercules in my fingers.

Walker had a kind of terror in his voice. "How do we know that all these boxes have gold in them? We never looked."

"I know," said Coertze. "I tested the weight of every box fifteen years ago. I made sure all right. There's about three tons of gold in this truck and another ton in the next one."

The gold had an insidious fascination and we were reluctant to leave it. For Walker and Coertze this was the culmination of the battle which was fought on that dusty road fifteen years previously. For me, it was the end of a tale that had been told many years before in the bars of Cape Town.

I suddenly pulled myself together. It was *not* the end of the tale, and if we wanted the tale to have a happy ending there was still much to do.

"O.K., let's break it up," I said. "There's a lot more to see and a hell of a lot to do."

The golden spell broken, we went to the next truck and Coertze again hammered the tailboard free. The bullion boxes were hidden this time, lying on the floor of the truck with other boxes piled on top.

"That's the box with the crown in it," said Walker excitedly.

We all climbed in, squashed at the back of the truck, and Coertze looked round. He suddenly glanced at Francesca and said, "Open that box and take your pick." He pointed to a stout case with a broken lock.

She opened the box and gasped. There was a shimmer of coruscating light, the pure white of diamonds, the bright green of emeralds and the dull red of rubies. She stretched forth her hand and picked out the first thing she encountered. It was a diamond and emerald necklace.

She ran it through her fingers. "How lovely!"

There was a catch in Piero's voice. "How much would that be worth?" he asked huskily.

"I don't know," I said. "Fifty thousand pounds, perhaps. That is, if the stones are real," I ended sardonically.

Coertze said, "Get this stuff out, then we can see what we have. I didn't have time when we put it in here."

"That's a good idea," I said. "But you won't have too much time now. It'll be dawn pretty soon, and we don't want to be seen round here."

We began to pull the boxes out. Coertze had thoughtfully left plenty of room between the trucks so it was easy enough. There were four boxes of jewellery, one filled with nothing else but wedding rings, thousands of them. I had a vague recollection that the patriotic women of Italy had given their wedding rings to the cause—and here they were.

There was the box containing the crown, a massive headpiece studded with jewels. There were eight large cases holding paper currency, neatly packeted and bound with rotting rubber bands. The lire had the original bank wrappers round each bundle. Then there were the remaining bullion boxes on the floor of the truck—another ton of gold.

Francesca went out to the car and brought in some flasks of coffee, and then we sat about examining the loot. The box from which Francesca had taken the necklace was the only one containing jewellery of any great value—but that was enough. I don't know anything about gems, but I conservatively estimated the value of that one box at well over a million pounds.

One of the other boxes was filled with various objects of value, usually in gold, such as pocket watches of bygone design, cigarette cases and lighters, gold medals and medallions, cigar cutters and all the other usual pieces of masculine jewellery. A lot of the pieces were engraved, but with differing names, and I thought that this must be the masculine equivalent of the wedding rings—sacrifices to the cause.

The third box contained the wedding rings and the last one was full of gold currency. There were a lot of British sovereigns and thousands of other coins which I identified as being similar to the coins shown to me by Aristide. There were American eagles and Austrian ducats and even some Tangier Hercules. That was a very heavy box.

Francesca picked up the necklace again. "Beautiful, isn't it?" I said.

"It's the loveliest thing I've seen," she breathed.

I took it from her fingers. "Turn round," I said, and fastened it round her neck. "This is the only opportunity you'll have of wearing it; it's a pity to waste it."

Her shoulders straightened and the triple line of diamonds

sparkled against her black sweater. Womanlike, she said, " Oh, I wish I had a mirror." Her fingers caressed the necklace.

Walker laughed and staggered to his feet, clutching the crown in both hands. He placed it on Coertze's head, driving the bullet head between the broad shoulders. " King Coertze," he cried hysterically. " All hail."

Coertze braced under the weight of the crown. " *Nee, man,*" he said, " I'm a Republican." He looked straight at me and smiled sardonically. " There's the king of this expedition."

To an outsider it would have been a mad sight. Four dishevelled and dirty men, one wearing a golden crown and with drying blood streaking his face, and a not-too-clean woman wearing a necklace worth a queen's ransom. We ourselves were oblivious to the incongruity of the scene; it had been with us too long in our imaginations.

I said, " Let's think of the next step."

Coertze lifted his hands and took off the crown. The fun was over; the serious work was to begin again.

" You'll have to finish off the entrance," I said. " That last bit isn't big enough to take the loot out."

Coertze said, " *Ja,* but that won't take long."

" Nevertheless, it had better be done now; it'll soon be dawn." I jerked my thumb at the third truck. " Anything of value back there?"

" There's nothing there but boxes of papers and dead Germans. But you can have a look if you want."

" I will," I said, and looked about the tunnel. " What I suggest is that Walker and I stay here to-day to get this stuff sorted out and moved to the front where it'll be easier to get out. It'll save time when the trucks come; I don't want them hanging about here for a long time."

I had thought out this move carefully. Coertze could be relied upon to keep a close watch on Piero and Francesca and would stand no nonsense from them when they went into Varsi.

But Coertze was immediately suspicious; he didn't want to leave me and Walker alone with the loot. I said, " Dammit, you'll seal us in, and even if we did make a break the stuff we could carry in our pockets wouldn't be worth worrying about compared with the rest of the treasure. All I want to do is save time."

After a glowering moment he accepted it, and he and Piero went to complete the entrance. I said to Walker, " Come on, let's take a look farther back."

He hesitated, and then said, " No. I'm not going back there. I'm not."

"I'll go with you," said Francesca quietly. "I'm not afraid of Germans, especially dead ones." She gave Walker a look of contempt.

I picked up the Tilley lamp and Walker said hysterically, "Don't take the light."

"Don't be a damn' fool," I said. "Take this to Coertze; it'll suit him better than a torch. You can give him a hand, too."

As he left I switched on my torch and Francesca did the same. I hefted the hammer and said, "O.K. Let's frighten all those ghosts."

The third truck was full of packing cases and weapons. There looked to be enough guns to start a war. I picked up a sub-machine-gun and cocked the action; it was stiff, but it worked and a round flew out of the breech. I thought that my gallant efforts at disarming Coertze and Piero were all wasted, or would have been if Coertze had remembered that all these guns were here. I wondered if the ammunition was still safe to use.

Francesca pushed some rifles aside and pulled the lid off one of the cases. It was full of files—dusty files with the *fasces* of the Fascist Government embossed on the covers. She pulled a file out and started to read, riffling the pages from time to time.

"Anything interesting?" I asked.

"It's about the invasion of Albania," she said. "Minutes of the meetings of the Army Staff." She took another file and became absorbed in it. "This is the same kind of thing, but it's the Ethiopian campaign."

I left her to the dusty records of forgotten wars and went back to the fourth truck. It was not pretty. The tunnel was very dry and apparently there had been no rats. The bodies were mummified, the faces blackened and the skin drawn tight into ghastly grins—the rictus of death. I counted the bodies—there were fifteen in the truck, piled in higgledy-piggledy like so many sides of beef—and two in the staff car, one of which was the body of an S.S. officer. There was a wooden case in the back of the truck but I did not investigate it—if it contained anything of value, the dead were welcome to keep it.

I went back to the staff car because I had seen something that interested me. Lying in the back, half hidden by the motor cycle, was a Schmeisser machine pistol. I picked it up and hefted it thoughtfully in my hand. I was thinking more of Coertze than of Metcalfe and my thoughts weren't pleasant. Coertze was suspected of having killed at least three men in order to get this treasure to himself. There was still the share-out to take place and it was on the cards that he would play the same game at some stage or other. The stake involved was tremendous.

The Schmeisser machine pistol is a very natty weapon which I had seen and admired during the war. It looks exactly like an ordinary automatic pistol and can be used as such, but there is a simple shoulder rest which fits into the holster and which clips into place at the back of the hand-grip so that you can steady the gun at your shoulder.

In principle, this is very much like the old Mauser pistol, but there the resemblance ends. Magazines for the Schmeisser come in two sizes—one of eight rounds like an ordinary pistol clip—and the long magazine holding about thirty rounds. With the long magazine in place and the gun switched to rapid fire you have a very handy sub-machine-gun, most effective at close range.

I had not fired a gun since the war and the thought of something which would make up for my lack of marksmanship by its ability to squirt out bullets was very appealing. I looked round to see if there were any spare clips but I didn't see any. Machine pistols were usually issued to sergeants and junior officers, so I prepared myself for an unpleasant task.

Then minutes later I had got what I wanted. I had the holster and belt, stiff with neglect, but containing the shoulder rest, four long clips and four short clips. There was another machine pistol, but I left that. I put the gun in the holster and left it resting in a niche in the tunnel wall together with the clips of ammunition. Then I went back to Francesca.

She was still reading the files by the light of her torch. I said, " Still reading history?"

She looked up. " It's a pitiful record; all the arguments and quarrels in high places, neatly tabulated and set down." She shook her head. " It is best that these files stay here. All this should be forgotten."

" It's worth a million dollars," I said, " if we could find an American university dishonest enough to buy it. Any historian would give his right arm for that lot. But you're right; we can't let it into the world outside—that would really give the game away."

" What is it like back there?" she asked.

" Nasty."

" I would like to see," she said and jumped down from the truck. I remembered the little girl of the war years who hated Germans, and didn't try to stop her.

She came back within five minutes, her face pale and her eyes stony, and would not speak of it. A long time afterwards she told me that she had vomited back there in sheer horror at the sight. She thought that the bodies ought to have been given decent burial, even though they were German.

When we got back to the front of the tunnel Coertze had finished his work and the entrance was now big enough to push the cases through. I sent Walker and Francesca back to the caravan to bring up food and bedding, then I took Coertze to one side, speaking in English so that Piero wouldn't understand.

"Is there any way to this mine other than by the road we came?" I asked.

"Not unless you travel cross-country," he replied.

I said, "You'll stay with Piero and Francesca at the caravan until late afternoon. You'll be able to see if anyone goes up the road; if anyone does you'll have to cut across country damn' quick and warn us, because we may be making a noise here. We'll probably sleep in the afternoon, so it should be all right then."

"That sounds fair enough," he said.

"Piero will probably start to look for those ammunition clips I threw away," I said. "So you'll have to keep an eye on him. And when you go to Varsi to pick up the trucks, make sure that you all stick together and don't let them talk to anyone unless you're there."

"*Moenie panik nie,*" he said. "They won't slip anything over on me."

"Good," I said. "I'm just going to slip out for a breath of fresh air. It'll be the last I'll get for a long time."

I went outside and strolled about for a while. I thought that everything was going well and if it stayed that way I would be thankful. Only one thing was worrying me. By bringing Francesca and Piero with us, we had cut ourselves off from our intelligence service and we didn't know what Metcalfe and Torloni were up to. It couldn't be helped, but it was worrying all the same.

After a while Piero came from the tunnel and joined me. He looked at the sky and said, "It will soon be dawn."

"Yes," I said. "I wish Walker and Francesca would come back," I turned to him. "Piero, something is worrying me."

"What is it, Signor Halloran?"

I said, "Coertze! He still has his gun, and I think he will try to look for those ammunition clips I threw away."

Piero laughed. " I will watch him. He will not get out of my sight."

And that was that. Those two would be so busy watching each other that they wouldn't have time to get up to mischief, and they would stay awake to watch the road. I rather fancied myself as a Machiavelli. I was no longer worrying too much about Francesca; I didn't think she would double-cross anyone. Piero was different; as he had said himself—gold has a bad effect on the character.

A few minutes later, Walker and Francesca came back in the car bringing food and blankets and some upholstered cushions from the caravan to use as pillows. I asked Walker discreetly, " Any trouble?"

" Nothing," he said.

The first faint light of morning was in the east. I said, " Time to go in," and Walker and I went back into the tunnel. Coertze began to seal up the entrance and I helped him from the inside. As the wall of rock grew higher I began to feel like a medieval hermit being walled up for the good of his soul. Before the last rocks were put in place Coertze said, " Don't worry about Varsi, it will be all right."

I said, " I'll be expecting you to-morrow at nightfall."

" We'll be here," he said. " You don't think I trust you indefinitely with all that stuff in there?"

Then the last rock sealed the entrance, but I heard him scuffling about for a long time as he endeavoured to make sure that it looked normal from the outside.

I went back into the tunnel to find Walker elbow deep in sovereigns. He was kneeling at the box, dipping his hands into it and letting the coins fall with a pleasant jingling sound. " We might as well make a start," I said. " We'll get half of the stuff to the front, then have breakfast, then shift the other half. After that we'll be ready for sleep."

The job had to be done so we might as well do it. Besides, I wanted to get Walker dead tired so that he would be heavily asleep, when I went to retrieve the Schmeisser.

The first thing we did was to clear the fallen rocks from in front of the first truck. This would be our working space when we had to disguise the bullion boxes and recrate the other stuff. We worked quickly without chatting. There was no sound except our heavy breathing, the subdued roar of the Tilley lamp and the ocasional clatter of a rock.

After an hour we had a clear space and began to bring the gold to the front. Those bullion boxes were damnably heavy and needed careful handling. One of them nearly fell on Walker's foot before I evolved the method of letting them drop from the lorry on to the piled-up caravan cushions. The cushions suffered but that was better than a broken foot.

It was awkward getting them to the front of the tunnel. The space between the lorry and the wall was too narrow for the two of us to carry a box together and the boxes were a little too heavy for one man to carry himself. I swore at Coertze for having reversed the trucks into the tunnel.

Eventually I hunted round among the trucks and found a long towing chain which we fastened round each box in turn so that we could pull it along the ground. The work went faster then.

After we had emptied all the gold from the first truck and had taken it to the front, I declared a breakfast break. Francesca had prepared a hot meal and there was plenty of coffee. As we ate we conversed desultorily.

"What will you do with your share?" I asked Walker curiously.

"Oh, I don't know," he said. "I haven't any real plans. I'll have a hell of a good time, I'll tell you that."

I grimaced. The bookmakers would take a lot of it, I guessed, and the distilleries would show a sudden burst in their profits for the first year, and then Walker would probably be dead of cirrhosis of the liver and delirium tremens.

"I'll probably do a lot of travelling," he said. "I've always wanted to travel. What will you do?"

I leaned my head back dreamily. "Half a million is a lot of money," I said. "I'd like to design lots of boats, the experimental kind that no one in their right minds would touch with a barge pole. A big cruising catamaran, for instance; there's a lot of work to be done in that field. I'd have enough money to have any design tank-tested as it should be done. I might even finance a private entry for the America's Cup—I've always wanted to design a 12-metre, and wouldn't it be a hell of a thing if my boat won?"

"You mean you'd go on *working*?" said Walker in horror.

"I like it," I said. "It's not work if you like it."

And so we planned our futures, going from vision to wilder vision until I looked at my watch and said, "Let's get cracking; the sooner we finish, the sooner we can sleep." It was nine o'clock and I reckoned we would be through by midday.

We moved the gold from the third truck. This was a longer haul and so took more time. After that it was easy and soon there was nothing left except the boxes of paper currency. Walker looked at them and said hesitantly, " Shouldn't we . . .?"

" Nothing doing," I said sharply. " I'd burn the lot if I was sure no one would see the smoke."

He seemed troubled at the heresy of someone wanting to burn money and set himself to count it while I got my blankets together and prepared for sleep. As I lay down, he said suddenly, " There's about a thousand million lire here—that's a hell of a lot of money. And there's any amount of sterling. Thousands of British fivers."

I yawned. " What colour are they?"

" White," he said. " The biggest notes I've seen."

" You pass one of those and you're for the high jump," I said. " They changed the design of the fiver when they discovered that the Germans had forged God knows how many millions. Come to think of it, it's quite likely that those are of German manufacture."

He seemed disappointed at that, and I said, " Get some sleep; you'll be glad of it later."

He gathered his blankets and settled himself down. I lay awake, fighting off sleep, until I heard the slow, regular breathing of deep slumber, then I got up and softly made my way down the tunnel. I retrieved the Schmeisser and the clips and brought them back. I didn't know where to put them at first, then I found that the cushion I was using as a pillow was torn and leaking stuffing. I tore out some more of the stuffing and put the gun and the clips inside. It made a hard pillow, but I didn't mind that—if people were going to wave guns at me, I wanted one to wave back.

IV

Neither of us slept very well—we had too much on our minds. I lay, turning restlessly, and hearing Walker doing the same until, at last, we could stand it no longer and we abandoned the pretence of sleep. It was four in the afternoon and I reckoned that the others should be starting for Varsi just about then.

We went up to the front of the tunnel and checked everything again, then settled to wait for nightfall. It could have been night then, if my watch hadn't told us otherwise, because there was no light in the tunnel except for the bright circle cast by the lamp, which quickly faded into darkness.

Walker was nervous. Twice he asked me if I heard a noise, not from the entrance but from back in the tunnel. The bodies of the men he had killed were worrying him. I told him to go back and look at them, thinking the shock treatment might do him good, but he refused to go.

At last I heard a faint noise from the entrance. I took the hammer in my hand and waited—this might not be Coertze at all. A rock clattered and a voice said, "Halloran?"

I relaxed and blew my cheeks out. It *was* Coertze.

Another rock clattered and I said, "Is everything all right?"

"No trouble at all," he said, furiously pulling down the screen of rocks. "The trucks are here."

Walker and I helped to push down the wall from the inside and Coertze shone a torch in my face. "Man," he said. "But you need a clean-up, ay."

I could imagine what I looked like. We had no water for washing and the dust lay heavily upon us. Francesca stood next to Coertze. "Are you all right, Mr. Halloran?"

"I'm O.K. Where are the trucks?"

She moved, barely distinguishable in the darkness. "They are back there."

"There are four Italians," said Coertze.

"Do they know what they are doing here?" I asked swiftly.

Piero loomed up. "They know that this is secret, and therefore certainly illegal," he said. "But otherwise they know nothing."

I thought about that. "Tell two of them to go down to the caravan, strike camp, and then wait there. Tell them to keep a watch on the road and to warn us if anyone comes up. The other two must go into the hills overlooking the mine, one to the left, the other to the right. They must watch for anyone coming across counry. This is the tricky part and we don't want anyone surprising us when the gold is in the open."

Piero moved away and I heard him giving quick instructions. I said, "The rest of us will start work inside. Bring the timber from the trucks."

The trucks were all right, bigger than we needed. One of them was loaded with lengths of rough boxwood and there were also some crude crates that would do for putting the loose stuff in. We hauled out the wood and took it into the tunnel, together with the tools—a couple of saws, four hammers and several packets of nails—and we started to nail covers on to the bullion boxes, changing their shape and character.

With four of us it went quickly and, as we worked, we developed an assembly-line technique. Walker sawed the wood into the correct lengths, Coertze nailed on the bottoms and the tops, I put on the sides and Piero put on the ends. Francesca was busy transferring the jewels and the gold trivia from the original boxes into the crates.

Within three hours we had finished and all there was left to do was to take the boxes outside and load them into the trucks.

I rolled my blankets and took my pillow outside and thrust them behind the driving-seat of one of the trucks—that disposed of the Schmeisser very nicely.

The boxes were heavy but Coertze and Piero had the muscle to hoist them vertically into the trucks and to stow them neatly. Walker and I used the chain again to pull the boxes through the narrow entrance. Francesca produced some flasks of coffee and a pile of cut sandwiches and we ate and drank while we worked. She certainly believed in feeding the inner man.

At last we were finished. I said, "Now we must take away from the tunnel everything we have brought here. We mustn't leave a scrap of evidence that we have been here, not a thing that can be traced back to us."

So we all went back into the tunnel and collected everything—blankets, cushions, tools, torches, flasks, even the discarded bent nails and the fragments of stuffing from the torn cushions. All this went outside to be stowed in the trucks and I stayed behind to take one last look round. I picked up a length of wood that had been forgotten and turned to leave.

Then it happened.

Coertze must have been hasty in shoring up the last bit of the entrance—he had seen the gold and his mind wasn't on his job. As I turned to leave, the piece of timber I was carrying struck the side of the entrance and dislodged a rock. There was a warning creak and I started to run—but it was too late.

I felt a heavy blow on my shoulder which drove me to my knees. There was a rumble of falling rock and then I knew no more.

V

I came round fuzzily, hearing a voice, "Halloran, are you all right? Halloran!"

Something soft touched my cheek and then something cold and wet. I groaned and opened my eyes but everything was hazy. The back of my head throbbed and waves of pain washed forward into my eyes.

I must have passed out again, but the next time I opened

my eyes things were clearer. I heard Coertze saying, "Can you move your legs, man; can you move your legs?"

I tried. I didn't understand why I should move my legs but I tried. They seemed to move all right so, dizzily, I tried to get up. I couldn't! There was a weight on my back holding me down.

Coertze said, "Man, now, take it easy. We'll get you out of there, ay."

He seemed to move away and then I heard Francesca's voice. "Halloran, you must stay quiet and not move. Can you hear me?"

"I can hear you," I mumbled. "What happened?" I found it difficult to speak because the right side of my face was lying on something rough and hard.

"You are pinned down by a lot of rock," she said. "Can you move your legs?"

"Yes, I can move my legs."

She went away and I could hear her talking to someone. My wits were coming back and I realised that I was lying prone with a heavy weight on my back and my head turned so that my right cheek was lying on rock. My right arm was by my side and I couldn't move it; my left arm was raised, but it seemed to be wedged tight.

Francesca came back and said, "Now, you must listen carefully. Coertze says that if your legs are free then you are only held in your middle. He is going to get you out, but it wil be very slow and you mustn't move. Do you understand?"

"I understand," I said.

"How do you feel? Is there pain?" Her voice was low and gentle.

"I feel sort of numb," I said. "All I feel is a lot of pressure on my back."

"I've got some brandy. Would you like some?"

I tried to shake my head and found it impossible. "No," I said. "Tell Coertze to get cracking."

She went away and Coertze came back. "Man," he said. "You're in a spot, ay. But not to worry, I've done this sort of thing before. All you have to do is keep still."

He moved and then I heard the scrape of rock and there was a scattering of dust on my face.

It took a long time. Coertze worked slowly and carefully, removing rocks one at a time, testing each one before he took it away. Sometimes he would go away and I would hear a low-voiced conversation, but he always came back to work again with a slow patience.

At last he said, " It won't be long now."

He suddenly started to shovel away rocks with more energy and the weight on my back eased. It was a wonderful feeling. He said, " I'm going to pull you out now. It might hurt a bit."

" Pull away," I said.

He grasped my left arm and tugged. I moved. Within two minutes I was in the open air looking at the fading stars. I tried to get up, but Francesca said, " Lie still."

Dawn was breaking and there was enough light to see her face as she bent over me. The winged eyebrows were drawn down in a frown as her hands pressed gently on my body testing for broken bones. " Can you turn over?" she asked.

It hurt, but I turned on to my stomach and heard the rip as she cut away my shirt. Then I heard the sudden hiss of her breath. " Your back is lacerated badly," she said.

I could guess how badly. Her hands were soft and gentle as they moved over my back. " You haven't broken anything," she said in wonderment.

I grinned. To me it felt as though my back was broken and someone had built a fire on it, but to hear that there were no broken bones was good. She tore some cloth and began to bind the wounds and when she had finished I sat up.

Coertze held out a baulk of six-by-six. " You were damned lucky, man. This was across your back and kept the full weight of the rock off you."

I said, " Thanks, Kobus."

He coloured self-consciously and looked away. " That's all right—Hal," he said. It was the first time he had ever called me Hal.

He looked at the sky. " We had better move now." He appealed to Francesca. " Can he move?"

I got to my feet slowly. " Of course I can move," I said. Francesca made a sudden gesture which I ignored. " We've got to get out of here."

I looked at the tunnel. " You'd better bring down the rest of that little lot and make a good job of it. Then we'll leave."

Coertze went off towards the tunnel, and I said, " Where's Walker?"

Piero said, " He is sitting in a truck."

" Send him down to the caravan, and whistle up your other two boys—they can go with him. They can all leave now for Rapallo."

Piero nodded and went away. Francesca said, " Hadn't you better rest a little?"

"I can rest in Rapallo. Can you drive one of those?" I nodded towards a truck.

"Of course."

"Good. Coertze and Piero can take one; we'll take the other. I might not be able to manage the driving part, though."

I didn't want Piero and Francesca alone, and I wanted Walker to keep a watch on the other Italians. Of course, I could have gone as passenger with Piero, but if he tried anything rough I was no match for him in my beat-up condition. Coertze could cope with him—so that left me with Francesca.

"I can manage," she said.

There was a rumble from the tunnel as Coertze pulled in the entrance, sealing it for ever, I hoped. He came back and I said, "You go with Piero in that truck; he'll be back in a minute. And don't tail me too close; we don't want to look like a convoy."

He said, "Think you'll be all right?"

"I'll be O.K." I said, and walked stiffly towards the truck in which I had left my gear. It was a painful business getting into the cab, but I managed in the end and rested gingerly in the seat, not daring to lean back. Francesca swung easily into the driving seat and slammed the door. She looked at me and I waved my hand. "Off we go."

She started the engine and got off badly by grinding the gears, and we went bouncing down the road from the mine, the rising sun shining through the windscreen.

The journey back to Rapallo was no joy-ride for me. The truck was uncomfortable as only trucks can be at the best of times, and for me it was purgatory because I was unable to lean back in the seat. I was very tired, my limbs were sore and aching, and my back was raw. Altogether I was not feeling too bright.

Although Francesca had said that she could drive the truck, she was not doing too well. She was used to the synchromesh gears of a private car and had a lot of trouble in changing the gears of the truck. To take my mind off my troubles we slowed down and I taught her how to double-declutch and after that things went easier and we began to talk.

She said, "You will need a doctor, Mr. Halloran."

"My friends call me Hal," I said.

She glanced at me and raised her eyebrows. "Am I a friend now?"

"You didn't kick me in the teeth when I was stuck in the tunnel," I said. "So you're my friend."

She slanted her eyes at me. "Neither did Coertze."

"He still needs me. He can't get the gold out of Italy without me."

"He *was* very perturbed," she agreed. "But I don't think he had the gold on his mind." She paused while she negotiated a bend. "Walker had the gold on his mind, though. He sat in a truck all the time, ready to drive away quickly. A contemptible little man."

I was too bemused by my tiredness to take in the implications of all this. I sat watching the ribbon of road unroll and I lapsed into an almost hypnotic condition. One of the things which fleetingly passed through my mind was that I hadn't seen the cigarette case which Walker had spoken of many years previously—the cigarette case which Hitler was supposed to have presented to Mussolini at the Brenner Pass in 1940.

I thought of the cigarette case once and then it passed from my mind, not to return until it was too late to do anything about it.

VI. METCALFE

The next day I felt better.

Everybody had got back to Palmerini's boatyard without untoward happenstance and we had moved into the big shed that was reserved for us. The trucks had been unloaded and returned to their owners with thanks, and the caravan stayed in a corner to provide cooking and sleeping space

But I was in no shape to do much work, so Walker and Coertze went to bring *Sanford* from the yacht basin, after I had checked on Metcalfe and Torloni. Francesca spoke to Palmerini and soon a procession of Italians slipped into the yard to make their reports. They spoke seriously to Francesca and ducked out again, obviously delighting in their return to the role of partisans.

When she had absorbed all they could tell her, Francesca came to me with a set face. "Luigi is in hospital," she said unhappily. "They broke his skull."

Poor Luigi. Torloni's men had not bothered to bribe him, after all. The harbour police were searching for the assailants but had had no success; and they wanted to see me to find out what had been stolen. As far as they were concerned it was just another robbery.

Francesca had an icy coldness about her. "We know who

they were," she said. "They will not walk out of Rapallo on their own legs."

"No," I said. "Leave them alone." I didn't want to show my hand yet because, with any luck, Metcalfe and Torloni might have fallen for the story I had planted. And for some reason, not yet clearly defined in my mind, I didn't want Francesca openly associated with us—she would still have to live in Italy when we had gone.

"Don't touch them," I said. "We'll take care of them later What about Metcalfe and Torloni?"

They were still in Genoa and saw each other every day. When they had found out that we had disappeared from Rapallo they had rushed up another three men, making five in all. Metcalfe had pulled the Fairmile from the water and Krupke was busy repainting the bottom. The Arab, Moulay Idriss, had vanished; no one knew where he was, but he was certainly not in Rapallo.

That all seemed satisfactory—except for the reinforcement of Torloni's men in Rapallo. I called Coertze and told him what was happening. "When you go to get *Sanford* tell the police that I've had a climbing accident, and that I'm indisposed. Make a hell of a fuss about the burglary, just as though you were an honest man. Go to the hospital, see Luigi and tell him that his hospital bill will be paid and that he'll get something extra for damages."

Coertze said, "Let me *donner* those bastards. They needn't have hit that old man."

"Don't go near them," I said. "I'll let you loose later, just before we sail."

He grumbled but held still, and he and Walker went to see what damage had been done to *Sanford*. After they had gone I had a talk with Piero. "You heard about Luigi?"

He pulled down his mouth. "Yes, a bad business—but just like Torloni."

I said, "I am thinking we might need some protection here."

"That is taken care of," he said. "We are well guarded."

"Does Francesca know about this?"

He shook his head. "Women do not know how to do these things—I will tell Madame when it is necessary. But this boatyard is well guarded; I can call on ten men within fifteen minutes."

"They'll have to be strong and tough men to fight Torloni's gangsters."

His face cracked into a grim smile. "Torloni's men know

nothing," he said contemptuously. "The men I have called are fighting men; men who have killed armed Germans with their bare hands. I would feel sorry for Torloni's gang were it not for Luigi."

I felt satisfied at that. I could imagine the sort of dock rats Torloni would have working for him; they wouldn't stand a chance again disciplined men accustomed to military tactics.

I said, "Remember, we want no killing."

"There will be no killing if they do not start it first. After that . . . ?" He shrugged. "I cannot be responsible for the temper of the men."

I left him and went into the caravan to clean and oil the Schmeisser. The tunnel had been dry and the gun hadn't taken much harm. I was more dubious about the ammunition; wondering if the charges behind the bullets had suffered chemical deterioration over the past fifteen years. That was something I would find out when the shooting started.

But perhaps there would be no shooting. There was a fair chance that Metcalfe and Torloni knew nothing of our connection with the partisans—I had worked hard enough to cover it. If Torloni attacked he would get the surprise of his life, but I hoped he wouldn't—I didn't want the Italians involved too much.

Coertze and Walker brought *Sanford* to the yard in the late afternoon and Palmerini's sons got busy slipping her and unstepping the mast. Coertze said, "We were followed by a fast launch."

"So they know we are here?"

"*Ja*," he said. "But we made them uncomfortable."

Walker said, "We took her out, and they had to follow us because they thought we were leaving. There was a bit of a lop outside the harbour and they were sea-sick—all three of them." He grinned. "So was Coertze."

"Did they do much damage to *Sanford* when they broke in?"

"Not much," replied Coertze. "They turned everything out of the lockers, but the police had cleaned up after the pigs."

"The furnaces?"

"All right; those were the first things I checked."

That was a relief. The furnaces were now the king-pins of the plan and if they had gone the whole of our labour would have been wasted. There would have been no time to replace them and still meet the deadline at Tangier. As it was, we would have to work fast.

Coertze got busy getting the furnaces out of *Sanford*. It wasn't a long job and soon he was assembling them on a bench in the corner of the shed. Piero looked at them uncomprehendingly but said nothing.

I realised it would be pointless to try to conceal our plan from him and Francesca—it just couldn't be done. And in any case, I was getting a bit tired of the shroud of suspicion with which I had cloaked myself. The Italians had played fair with us so far and we were entirely at their mercy, anyway; they could take the lot any time they wanted if they felt so inclined.

I said, "We're going to cast a new keel for *Sanford*."

Piero said, "Why? What is wrong with that one?"

"Nothing, except it's made of lead. I'm a particular man—I want a keel of gold."

His face lit up in a delighted smile. "I wondered how you were going to get the gold out of the country. I thought about it and could see no way, but you seemed so sure."

"Well, that's how we're going to do it," I said, and went over to Coertze. "Look," I said. "I'm not going to be good for any heavy work over the next few days. I'll assemble these gadgets—it's a sitting job—you'd better be doing something else. What about the mould?"

"I'll get started on that," he said. "Palmerini has plenty of moulding sand."

I unfastened my belt and, from the hidden pocket, I took the plan of the new keel I had designed many months previously. I said, "I had Harry make the alterations to the keelson to go with the new keel. He thought I was nuts. All you've got to do is to cast the keel to this pattern and it'll fit sweetly."

He took the drawing and went off to see Palmerini. I started to assemble the furnaces—it wasn't a long job and I finished that night.

II

I suppose that few people have had occasion to cut up gold ingots with a hacksaw. It's a devilish job because the metal is soft and the teeth of the saw blades soon become clogged. Walker said it was like sawing through treacle.

It had to be done because we could only melt a couple of pounds of gold at a time, and it was Walker's job to cut up the ingots into nice handy pieces. The gold dust was a problem which I solved by sending out for a small vacuum cleaner which Walker used assiduously, sucking up every particle of gold he could find.

And when he had finished sawing for the day he would sweep round his bench and wash the dust in a pan just like an old-time prospector. Even with all those precautions I reckon we must have wasted several pounds of gold in the sawing operation.

We all gathered round to watch the first melt. Coertze dropped the small piece of gold on to the graphite mat and switched on the machine. There was an intense white flare as the mat went incandescent and the gold drooped and flowed and, within seconds, was ready for pouring into the mould.

The three furnaces worked perfectly but as they were only laboratory instruments after all, and could only take a small amount at a time, it was going to be a long job. Inside the mould we put a tangle of wires which was to hold the gold together. Coertze was dubious about the method of pouring so little at a time and several times he stopped and removed gold already poured.

"This keel will be so full of faults and cracks I don't think it'll hold," he said.

So we put in more and more wires and poured the gold round them, hoping they would bind the mass together.

I was stiff and sore and to bend was an agony, so there was not much I could do to help effectively. I discussed this with Coertze, and said, "You know, one of us had better show his face in Rapallo. Metcalfe knows we're here and it'll look odd if we all stay in this shed and never come out. He'll know we're up to something."

"You'd better wander round town then," said Coertze. "You can't do much here."

So after Francesca had rebandaged my back, I went into town and up to the Yacht Club. The secretary commiserated with me on the fact that *Sanford* had been broken into and hoped that nothing had been stolen. "It cannot have been done by men of Rapallo," he said. "We are very strict about that here."

He also looked at my battered face in mute inquiry, so I smiled and said, "Your Italian mountains seem to be made of harder rock than those in South Africa."

"Ah, you've been climbing?"

"Trying to," I said. "Allow me to buy you a drink."

He declined, so I went into the bar and ordered a Scotch, taking it to the table by the window where I could look over the yacht basin. There was a new boat in, a large motor yacht of about a hundred tons. You see many of those in the Mediterranean—the luxury boats of the wealthy. They put to sea in the calmest of weather and the large paid crews have the life of Reilly—hardly any work and plenty of shore time. Idly, I focused the club binoculars on her. Her name was *Calabria.*

When I left the club I spotted my watchers and took delight in leading them to innocent places which any tourist might have visited. If I had been fitter I would have walked their legs off, but I compromised by taking a taxi. Their staff-work was good, because I noticed a cruising car come up from nowhere and pick them up smoothly.

I went back and reported to Francesca. She said, "Torloni has sent more men into Rapallo."

That sounded bad. "How many?"

"Three more—that makes eight. We think that he wants enough men to follow each of you, even if you split up. Besides, they must sleep sometimes, too."

"Where's Metcalfe?"

"Still in Genoa. His boat was put into the water this morning."

"Thanks, Francesca, you're doing all right," I said.

"I will be glad when this business is finished," she said sombrely. "I wish I had never started it."

"Getting cold feet?"

"I do not understand what you mean by that; but I am afraid there will be much violence soon."

"I don't like it, either," I said candidly. "But the thing is under way; we can't stop now. You Italians have a phrase for it—*che sera, sera.*"

She sighed. "Yes, in a matter like this there is no turning back once you have begun."

I left her sitting in the caravan, thinking that she was beginning to realise that this was no light-hearted adventure she had embarked upon. This was deadly serious, a game for high stakes in which a few murders would not be boggled at, at least, not by the opposition—and I wasn't too sure about Coertze.

The keel seemed to be going well. Coertze and Piero were sweating over the hot furnaces, looking demoniacal in the sudden bursts of light. Coertze pushed up his goggles and said, "How many graphite mats did we have?"

"Why?"

"They don't last long. I'm not getting more than four

melts out of each, then they burn out. We might run out of mats before thc job's finished."

"I'll check on it," I said, and went to figure with pencil and paper. After checking my calculations and recounting the stock of mats I went back to Coertze. "Can you squeeze five melts out of a mat?"

He grunted. "We'll have to be careful about it, which means we'll be slower. Can we afford the time?"

"If we burn out the mats before the job's done then the time won't matter—it'll be wasted anyway. We'll have to afford the time. How many melts a day can you do at five melts to a mat?"

He thought about that. "It'll cut us down to twelve melts an hour, no more than that."

I went away to do some more figuring. Taking the gold at 9000 pounds, that meant 4,500 melts of which Coertze had already done 500. Twelve melts an hour meant 340 working hours—at twelve hours a day, twenty-eight days.

Too long—start again.

Three hundred and forty hours working at sixteen hours a day—twenty-one days. But could he work sixteen hours a day? I cursed my lacerated back which kept me from helping, but if anything happened and it got worse then I was sure the plan would be torpedoed. Somebody had to take *Sanford* out and I had an increasing distrust of Walker, who had grown silent and secretive.

I went back to Coertze, walking stiffly because my back was hurting like hell. "You'll have to work long hours," I said. "Time's running out."

"I'd work twenty-four hours a day if I could," he said. "But I can't, so I'll work till I drop."

I thought maybe I'd better go at it at different way, so I stood back and watched how Coertze and Piero were going about the job. Soon I had ideas about speeding it up.

The next morning I took charge. I told Coertze to do nothing but pour gold; he must not have anything to do with loading the furnaces or cleaning mats—all he had to do was pour gold. Piero I assigned to melting the gold and to passing the furnace with the molten gold to Coertze. The furnaces were light enough to be moved about so I arranged a table so that they could move bodily along it.

Walker had sawn plenty of gold, so I pulled him from his bench. He had to take a furnace from Coertze, replace the mat with a new one and put a chunk of gold on it ready for melting. Myself I set to the task of cleaning the used mats ready for re-use—this I could do sitting down.

All in all, it was a simple problem in time and motion study and assembly line technique. By the end of the day we were doing sixteen melts an hour without too many burnt-out mats.

So the days went by. We started by working sixteen hours a day but we could not keep it up and gradually our daily output dropped in spite of the increase in the hourly output. Mistakes were made in increasing numbers and the percentage of burnt-out mats went up sharply. Working in those sudden bursts of heat from the furnaces was hellish; we all lost weight and our thirst was unquenchable.

When the output dropped below 150 melts a day with another 2000 to go I began to get really worried. I wanted a clear three weeks to sail to Tangier and it looked as though I was not going to get them.

Obviously something had to be done.

That evening, when we were eating supper after finishing work for the day, and before we turned exhaustedly into our berths, I said, "Look, we're too tired. We're going to have a day off, to-morrow. We do nothing at all—we just laze about."

I was taking a chance, gambling that the increased output by refreshed men would more than offset the loss of a day. But Coertze said bluntly. "No, we work. We haven't the time to waste."

Coertze was a good man if a bit bull-headed. I said, "I've been right up to now, haven't I?"

He grudgingly assented to that.

"The output will go up if we have a rest," I said. "I promise you."

He grumbled a little, but didn't press it—he was too tired to fight. The others agreed lacklustrely, and we turned in that night knowing that the next day would be a day of rest.

III

At breakfast, next morning, I asked Francesca, "What's the enemy doing?"

"Still watching."

"Any reinforcements?"

She shook her head, "No, there's just the eight of them. They take it in turns."

I said, "We might as well give them some exercise. We'll split up and run them about town, or even outside it. They've been having it too easy lately."

I looked at Coertze. "But don't touch them—we're not ready to force a showdown yet, and the later it comes the better for us. We can't afford for any one of us to be put out of action now; if that happens we're sunk. It'll take all our time to cast the keel and meet the deadline as it is."

To Walker I said, "And you keep off the booze. You might be tempted, but don't do it. Remember what I said in Tangier?"

He nodded sullenly and looked down at his plate. He had been too quiet lately to suit me and I wondered what he was thinking.

I said to Francesca, "I thought you were getting a jeweller to appraise the gems."

"I will see him to-day," she said. "He will probably come to-morrow."

"Well, when he comes, it must be in disguise or something. Once Torloni's men know that there are jewels involved there may be no holding them."

Piero said, "Palmerini will bring him hidden in a lorry."

"Good enough." I got up from the table and stretched. "Now to confuse the issue and the enemy. We'll all leave in different directions. Piero, you and Francesca had better leave later; we don't want any connection to be made between us. Will this place be safe with us all gone?"

Francesca said, "There'll be ten of our men in the yard all day."

"That's fine," I said. "Tell them not to be too conspicuous."

I felt fine as I walked into town. My back was healing and my face no longer looked like a battlefield. I was exhilarated at the prospect of a day off work and Coertze must have been feeling even better, I thought. He had not left Palmerini's yard since he had brought *Sanford* in, while I had had several visits to town.

I spent the morning idling, doing a little tourist shopping in the Piazza Cavour where I found a shop selling English books. Then I had a lengthy stay at a boulevard café where I leisurely read a novel over innumerable cups of coffee, something I had not had time for for many months.

Towards midday I went up to the Yacht Club for a drink. The bar seemed noisier than usual and I traced the disturbance to an argumentive and semi-drunken group at the far end of the room. Most members were pointedly ignoring this demonstration but there were raised eyebrows at the more raucous shouts. I ordered a Scotch from the steward and said, "Why the celebration?"

He sneered towards the end of the bar. "No celebration, signor; just idle drunkenness."

I wondered why the secretary didn't order the men from the club and said so. The steward lifted his shoulders helplessly. "What can one do, signor? There are some men who can break all rules—and here is one such man."

I didn't press it; it was no affair of mine and it wasn't my business to tell the Italians how to run the club in which I was their guest. But I did take my drink into the adjoining lounge where I settled down to finish the novel.

It was an interesting book, but I never did get it finished, and I've often wondered how the hero got out of the predicament in which the author placed him. I had not read half a dozen pages when a steward came up and said, "There is a lady to see you, signor."

I went into the foyer and saw Francesca. "What the devil are you doing here?" I demanded.

"Torloni is in Rapallo," she said.

I was going to speak when the club secretary came round the corner and saw us. I said, "You'd better comc inside; it's too damn' conspicuous here."

The secretary hurried over, saying, "Ah, Madame, we have not had the honour of a visit from you for a long time."

I was a member of the club—if only honorary—so I said, "Perhaps I could bring Madame into the club as my guest?"

He loked unaccountably startled and said nervously, "Yes, yes, of course. No, there is no need for Madame to sign the book."

As I escorted Francesca into the lounge I wondered what was agitating the secretary, but I had other things on my mind so I let it slide. I seated Francesca and said, "You'd better have a drink."

"Campari," she said, and then quickly, "Torloni brought a lot of men with him."

"Relax," I said, and ordered a Campari from the lounge steward. When he had left the table I said, "What about Metcalfe?"

"The Fairmile left Genoa; we don't know where it is."

"And Torloni? Where is he?"

"He booked into a hotel on the Piazza Cavour an hour ago."

That was when I had been sitting in the pavement café. I might even have seen him. I said, "You say he brought some men with him?"

"There are eight men with him."

That was bad; it looked as though an attack was building up. Eight plus eight made sixteen, plus Torloni himself and possibly Metcalfe, Krupke, the Moroccan and what other crew the Fairmile might have. More than twenty men!

She said, "We had to work quickly. There was a lot of reorganising to do—that is why I came here myself, there was no one else."

I said, "Just how many men have *we* got?"

She furrowed her brow. "Twenty-five—possibly more later. I cannot tell yet."

That sounded better; the odds were still in our favour. But I wondered about Torloni's massing of force. He would not need so many men to tackle three presumably unsuspecting victims, therefore he must have got wind of our partisan allies, so perhaps we wouldn't have the advantage of surprise.

The steward came with the Campari and as I paid him Francesca looked from the window over the yacht basin. When the steward had gone, she said, "What ship is that?"

"Which one?"

She indicated the motor yacht I had noticed on my earlier visit to the club. "Oh, that! It's just some rich man's floating brothel."

Her voice was strained. "What is the name?"

I hunted in my memory. "Er—*Calabria,* I think."

Her knuckles were clenched white as she gripped the arms of her chair. "It is Eduardo's boat," she said in a low voice.

"Who is Eduardo?"

"My husband."

A light dawned on me. So that was why the secretary had been so startled. It is not very usual for a stranger to ask a lady to be his guest when the lady's husband is within easy reach and possibly in the club at that very moment. I chuckled and said, "I'll bet he's the chap who is kicking up such a shindy in the bar."

She said, "I must go."

"Why?"

"I do not wish to meet him." She pushed her drink to one side and picked up her handbag.

I said, "You might as well finish your drink. It's the first drink I've ever bought you. No man is worth losing a drink over, anyway."

She relaxed and picked up the Campari. "Eduardo is not worth anything," she said tightly. "All right, I will be civilised and finish my drink; then I will go."

But we did meet him, after all. Only an Estrenoli—from what I had heard of the breed—would have paused dramatically in the doorway, veered over to our table and have addressed Francesca as he did.

"Ah, my loving wife," he said. "I'm surprised to find you here in civilised surroundings. I thought you drank in the gutters."

He was a stocky man, with good looks dissipated by red-veined eyes and a slack mouth. A wispy moustache disfigured his upper lip and his face was flushed with drink. He ignored me altogether.

Francesca looked stonily ahead, her lips compressed, and did not turn to face him even when he dropped heavily into a chair by her side.

I said, "You weren't invited to sit with us, signor."

He swung round and gave a short laugh, looking at me with an arrogant stare. He turned back to Francesca. "I see that even the Italian scum is not good enough for you now; you must take foreign lovers."

I stretched out my foot and hooked it behind the rung of his chair, then pulled hard. The chair slid from under him and he tumbled on to the floor and sprawled full length. I got up and stood over him. "I said you weren't invited to sit down."

He looked up at me, his face suffused with anger, and slowly scrambled to his feet. Then he glared at me. "I'll have you out of the country within twenty-four hours," he screamed. "Do you know who I am?"

The chance was too good to miss. "Scum usually floats on top," I said equably, then I hardened my voice. "Estrenoli, go back to Rome. Liguria isn't a healthy place for you."

"What do you mean by that?" he said uneasily. "Are you threatening me?"

"There are fifty men within a mile of here who would fight each other for the privilege of cutting your throat," I said. "I'll tell you what; I'll give *you* twenty-four hours to get out of Liguria. After that I wouldn't give a busted lira for your chances."

I turned to Francesca. "Let's get out of here; I don't like the smell."

She picked up her handbag and accompanied me to the door, walking proudly and leaving Estrenoli standing there impotently. I could hear a stifled buzz of comment in the lounge and there were a few titters at his discomfiture. I suppose there were many who had wanted to do the same thing but he was too powerful a man to cross. I didn't give a damn; I was boiling with rage.

The tittering was too much for Estrenoli and he caught up with us as we were crossing the foyer. I felt his hand on my shoulder and turned my head. "Take your hand off me," I said coldly.

He was almost incoherent in his rage. "I don't know who you are, but the British Ambassador will hear about this."

"The name's Halloran, and take your goddamm hand off me."

He didn't. Instead his hand tightened and he pulled me round to face him.

That was too much.

I sank three stiff fingers into his soft belly and he gasped and doubled up. Then I hit him with my fist as hard as I could. All the pent-up frustrations which had accumulated over the past weeks went into that blow; I was hitting Metcalfe and Torloni and all the thugs who were gathering like vultures. I must have broken Estrenoli's jaw and I certainly scraped my knuckles. He went down like a sack of meal and lay in a crumpled heap, blood welling from his mouth.

In the moment of hitting him I felt a fierce pain in my back. "Christ, my back!" I groaned, and turned to Francesca. But she was not there.

Instead, I was face to face with Metcalfe!

"What a punch!" he said admiringly. "That bloke's got a busted jaw for sure; I heard it go. Ever consider fighting light-heavyweight, Hal?"

I was too astounded to say anything, then I remembered Francesca and looked about wildly. She moved into sight from behind Metcalfe.

He said, "Wasn't this character saying something about the British Ambassador?" He looked about the foyer. Luckily it was deserted and no one had seen the fracas. Metcalfe looked at the nearest door, which was the entrance to the men's room. He grinned. "Shall we lug the guts into the neighbour room?"

I saw his point and together we dragged Estrenoli into the lavatory and stuffed him into a cubicle. Metcalfe straightened and said, "If this bird is on speaking terms with the British Ambassador he must be a pretty big noise. Who is he?"

I told him and Metcalfe whistled. "When you hit 'em, you

hit 'em big! Even I have heard of Estrenoli. What did you slug him for?"

"Personal reasons," I said.

"Connected with the lady?"

"His wife."

Metcalfe groaned. "Brother you do get complicated. You're in a jam, for sure—you'll be tossed out of Italy on your ear within twelve hours." He scratched behind his ear. "But maybe not; maybe I can fix it. Wait here and don't let anyone use this john. I'll tell your girl-friend to stick around—and I'll be back in a couple of minutes."

I leaned against the wall and tried to think coherently about Metcalfe, but I couldn't. My back was hurting like hell and there was a dull throbbing in the hand with which I had hit Estrenoli. It looked as though I had made a mess of everything. I had repeatedly warned Coertze not to get into brawls and now I was guilty of that same thing—and mixed up with Metcalfe to boot.

Metcalfe was as good as his word and was back within two minutes. With him was a squat, blue-jowled Italian dressed in a sharp suit. Metcalfe said, "This is a friend of mine, Guido Torloni. Guido, this is Peter Halloran."

Torloni looked at me in quick surprise. Metcalfe said, "Hal's in a jam. He's broken a governmental jaw." He took Torloni on one side and they spoke in low tones. I watched Torloni and thought that the mess was getting worse.

Metcalfe came back. "Don't worry, Guido can fix it, he can fix anything."

"Even Estrenoli?" I said incredulously.

Metcalfe smiled. "Even Estrenoli. Guido is Mr. Fixit himself in this part of Italy. Come, let's leave him to it."

We went into the foyer and I did not see Francesca. Metcalfe said, "Mrs. Estrenoli is waiting outside in my car."

We went out to the car and Francesca said, "Is everything all right?"

"Everything is fine," I said.

Metcalfe chuckled. "Excepting your husband, Madame. He will be very sorry for himself when he wakes up."

Francesca's hand was on the edge of the door. I put my hand over hers and pressed it warningly. "I'm sorry," I said. "Francesca, this is Mr. Metcalfe, an old friend of mine from South Africa."

I felt her fingers tense. I said quickly, "Mr. Metcalfe's friend, Mr. Torloni, is looking after your husband. I'm sure he'll be all right."

"Oh yes," agreed Metcalfe cheerfully. "Your husband will be fine. He won't make trouble for anyone." He suddenly frowned. "How's your back, Hal? You'd better have it seen to right away. If you like I'll drive you to a doctor."

"It doesn't matter," I said. I didn't want to be driving anywhere with Metcalfe.

"Nonsense!" he said. "Who is your doctor?"

It made a bit of a difference if he would take us to a doctor of *our* choice. I looked at Francesca who said, "I know a good doctor."

Metcalfe clapped his hands together. "Fine. Let's get cracking."

So he drove us through the town and Francesca pointed out a doctor's rooms. Metcalfe pulled up and said, "You two go in; I'll wait for you here and give you a lift to Palmerini's yard."

That was another facer. Apparently Metcalfe didn't mind us knowing that he knew our whereabouts. There was something queer in the air and I didn't like it.

As soon as we got into the doctor's waiting-room Francesca said, "Is *that* Metcalfe? He seems a nice man."

"He is," I said. "But don't get in his way or you'll get run over." I winced as my back gave a particularly nasty throb. "What the hell do we do now?"

"Nothing has changed," said Francesca practically. "We knew they would be coming. Now they are here."

That was true. I said, "I'm sorry I hit your husband."

"I'm not," she said simply. "The only thing I'm sorry for is that you got hurt doing it. And that it might cause trouble for you."

"It won't," I said grimly. "Not while he's in Torloni's hands. And that's another thing I don't understand—why should Metcalfe and Torloni be interested in getting me out of trouble? It doesn't make sense."

The doctor was ready for us then and he looked at my back. He said that I had torn a ligament and proceeded to truss me up like a chicken. He also bound up my hand, which was a bit damaged where the knuckles had been scraped on Estrenoli's teeth. When we came out Metcalfe waved at us from the car, and called, "I'll take you down to the yard."

There didn't seem to be much point in refusing under the circumstances so we climbed into the car. As we were pulling away I said casually, "How did you know we were in Palmerini's yard?"

"I knew you were cruising in these waters so I asked the

Port Captain if you'd shown up yet," said Metcalfe airily. "He told me all about you."

It was logical enough, and if I hadn't known better I might have believed him. He said, "I hear you're having trouble with your keel."

That was cutting a bit near the bone. I said, "Yes, I tried an experimental method of fastening but it doesn't seem to be working out. I might have to take the keel off and refasten it."

"Make a good job of it," he said. "It would be a pity if it dropped off when you're off-shore. You'd capsize immediately."

This was an uncomfortable conversation; it was reasonable small boat shop-talk, but with Metcalfe you never knew. To my relief he switched to something else. "What did you do to your face? Been in another brawl lately?"

"I fell off a mountain," I said lightly.

He made a sucking sound with his lips in commiseration. "You want to take more care of yourself, Hal, my boy. I wouldn't want anything to happen to you."

This was too much. "Why the sudden solicitude?" I asked acidly.

He turned in surprise. "I don't like seeing my friends get bashed about, especially you. You're quite a handsome feller, you know." He turned to Francesca. "Isn't he?"

"I think so," she said.

I was surprised at that. "I'll survive," I said, as Metcalfe drew up at the gate of the boatyard. "I'm getting to be an expert at it."

Francesca and I got out of the car, and Metcalfe said, "Not going to show me your new keel fastening, Hal?"

I grinned. "Hell, I'm a professional designer; I never show my mistakes to anyone." If he could play fast and loose in a hinting conversation, so could I.

He smiled. "Very wise of you. I'll be seeing you around, I suppose?"

I stepped up to the car out of earshot of Francesca. "What will happen to Estrenoli?"

"Nothing much. Guido will take him to a good, safe doctor and have him fixed up, then he'll dump him in Rome after throwing a hell of a scare into him. It's my guess that Estrenoli's not very brave and our Guido is a very scary character when he wants to be. There'll be no more trouble."

I stepped back from the car, relieved. I had been afraid that Estrenoli would be dumped at the bottom of the bay in a concrete overcoat, and I didn't want anyone's life on my conscience, not even his. I said, "Thanks. Yes, I'll be seeing you around. One can scarcely avoid it—in a town as small as this, can one?"

He put the car into gear and moved forward slowly, grinning from the side window. "You're a good chap, Hal; don't let anybody put one over on you."

Then he was gone and I was left wondering what the hell it was all about.

IV

The atmosphere in the shed was tense. As we walked through the yard I noticed that there were many more people about than usual; those would be Francesca's friends. When we got into the shed Piero strode up and said, "What happened at the club?" His voice was shaking with emotion.

"Nothing happened," I said. "Nothing serious." I saw a stranger in the background, a little man with bright, watchful eyes. "Who the devil's that?"

Piero turned. "That's Cariaceti, the jeweller—never mind him. What happened at the club? You went in and so did Madame; then this Metcalfe and Torloni went in; then you and Madame came out with Metcalfe. What is happening?"

I said, "Take it easy; everything is all right. We bumped into Estrenoli and he got flattened."

"Estrenoli?" said Piero in surprise, and looked at Francesca who nodded in confirmation. "Where is he now?" he demanded fiercely.

"Torloni's got him," I said.

That was too much for Piero. He sat on a trestle and gazed at the floor. "Torloni?" he said blankly. "What would Torloni want with Estrenoli?"

"Damned if I know," I said. "This whole thing is one of Metcalfe's devious plays. All I know is that I had a bust-up with Estrenoli and Metcalfe has removed him from circulation for a while—and don't ask me why."

He looked up. "It is said that you were very friendly with Metcalfe to-day." His voice was heavy with suspicion.

"Why not? There's nothing to be gained by antagonising him. If you want to know what happened, ask Francesca—she was there."

"Hal is right," said Francesca. "His treatment of Metcalfe was correct. He was given much provocation and refused to be annoyed by it. Besides," she said with a slight smile, "Metcalfe would seem to be a difficult man to hate."

"It is not difficult to hate Torloni," growled Piero. "And Metcalfe is his friend."

This wasn't getting us anywhere, so I said, "Where are Coertze and Walker?"

"In the town," said Piero. "We know where they are."

"I think they had better come in," I said. "Things may start to move fast—we'd better decide what to do next."

He silently got up and went outside. I walked over to the little jeweller. "Signor Cariaceti," I said. "I understand that you have come here to look at some gems."

"That is so," he said. "But I do not wish to remain here long."

I went back to Francesca. "You'd better turn Cariaceti loose among the jewels," I said. "There may not be much time."

She went to talk to Cariaceti and I looked moodily at the keel, still lacking nearly two tons of weight. Things were at a low ebb and I felt pretty desperate. It would take eight more days working at high pressure to finish the keel, another day to fasten it in position and another to replace the glass-fibre cladding and to launch *Sanford*.

Ten days! Would Metcalfe and Torloni wait that long?

After a little while Francesca came back. "Cariaceti is amazed," she said. "He is the happiest man I have ever seen."

"I'm glad someone is happy," I said gloomily. "This whole thing is on the point of falling to pieces."

She put her hand on my arm. "Don't blame yourself," she said. "No one could have done more than you."

I sat on the trestle. "I suppose things *could* be worse," I said. "Walker could get stinking drunk just when we need him, Coertze could run amok like a mad bull and I could fall and break a few bones."

She took my bandaged hand in hers. "I have never said this to any man," she said. "But you are a man I could admire very much."

I looked at her hand on mine. "Only admire?" I asked gently.

I looked up to see her face colouring. She took her hand away quickly and turned from me. "Sometimes you make me very annoyed, Mr. Halloran."

I stood up. "It was 'Hal' not very long ago. I told you that my friends called me Hal."

"I am your friend," she said slowly.

"Francesca, I would like you to be more than my friend," I said.

She was suddenly very still and I put my hand on her waist. I said, "I think I love you, Francesca."

She turned quickly, laughing through tears. "You only think so, Hal. Oh, you English are so cold and wary. I *know* that I love you."

Something seemed to give at the pit of my stomach and the whole dark shed suddenly seemed brighter. I said, "Yes, I love you; but I didn't know how to say it properly—I didn't know what you would say when I told you."

"I say 'bravo.'"

"We'll have a good life," I said. "The Cape is a wonderful place—and there is the whole world besides."

She saddened quickly. "I don't know, Hal; I don't know. I am still a married woman; I can't marry you."

"Italy isn't the world," I said softly. "In most other countries divorce is not dishonourable. The men who made the laws for divorce were wise men; they would never tie anyone to a man like Estrenoli for life."

She shook her head. "Here in Italy and in the eyes of my Church, divorce is a sin."

"Then Italy and your Church are wrong. I say it; even Piero says it."

She said slowly, "What is going to happen to my husband?"

"I don't know," I said. "Metcalfe tells me that he will be taken back to Rome—under escort."

"That is all? Torloni will not kill him?"

"I don't think so. Metcalfe said not—and I believe Metcalfe. He may be a scoundrel, but I've never caught him out in a black lie yet."

She nodded. "I believe him, too." She was silent for a while, then she said, "When I know that Eduardo is safe, then I will come away with you, to South Africa or any other place. I will get a foreign divorce and I will marry you, but Eduardo must be alive and well. I could not have that thing on my conscience."

I said, "I will see to it. I will see Metcalfe." I looked at the keel. "But I must also see this thing through. I have set my hand to it and there are others to consider—Coertze, Walker, Piero, all your men—I can't stop now. It isn't just the gold, you know."

"I know," she said. "You must have been hurt by someone to start a thing like this. It is not your natural way."

I said, " I had a wife who was killed by a drunkard like Walker."

" I know so little of your past life," he said in wonder. " I have so much to learn. Your wife—you loved her very much." It was not a question, it was a statement.

I told her a little about Jean and more about myself and for a while we talked about each other in soft voices, the way that lovers do.

V

Then Coertze came in.

He wanted to know what all the hurry was and why his rest day had been broken into. For a man who didn't want to stop work he was most averse to being interrupted in his brief pleasures.

I brought him up to date on events and he was as puzzled as any of us. " Why should Metcalfe want to help us?" he asked.

" I don't know, and I don't intend to ask him," I said. " He might tell me the truth and the truth might be worse than any suspicions we may have."

Coertze did as I had done and went to stare at the keel. I said, " Another eight days of casting—at the least."

" *Magtig,*" he burst out. " No one is going to take this away from me now." He took off his jacket. " We'll get busy right now."

" You'll have to do without me for an hour," I said. " I have an appointment."

Coertze stared at me but did not say anything as I struggled into my jacket. Francesca helped me to put it on over the carapace of bandaging under my shirt. " Where are you going?" she asked quietly.

" To see Metcalfe. I want to make things quite clear."

She nodded. " Be careful."

On the way out I bumped into Walker who looked depressed. " What's the matter with you?" I said. " You look as though you've lost a shilling and found a sixpence."

" Some bastard picked my pocket," he said savagely.

" Lose much?"

" I lost my ci . . ." He seemed to change his mind. " I lost my wallet."

" I wouldn't worry about that," I said. " We're going to lose the gold if we aren't careful. See Coertze, he'll tell you about it." I pushed past him, leaving him staring at me.

I went into Palmerini's office and asked if I could borrow his car. He didn't mind so I took his little Fiat and drove down to the yacht basin. I found the Fairmile quite easily and noted that it was not visible from the Yacht Club, which was why I hadn't spotted it earlier. Krupke was polishing the bright-work on the wheelhouse.

"Hi," he said. "Glad to see you. Metcalfe told me you were in town."

"Is he on board? I'd like to see him."

"Wait a minute," said Krupke and went below. He came back almost immediately. "He says you're to come below."

I jumped on to the deck and followed Krupke below to the main saloon. Metcalfe was lying on a divan reading a book. "What brings you here so soon?" he asked.

"I want to tell you something," I said, and glanced at Krupke.

"O.K., Krupke," said Metcalfe, and Krupke went out. Metcalfe opened a cupboard and produced a bottle and two glasses. "Drink?"

"Thanks," I said.

He poured out two stiff ones, and said, "Mud in your eye." We drank, then he said, "What's your trouble?"

"That story you told me about Torloni taking care of Estrenoli—is it true?"

"Sure. Estrenoli's with a doctor now."

"I just wanted to make sure," I said. "And to make certain, you can tell Torloni from me that if Estrenoli doesn't reach Rome safe and sound then I'll kill him personally."

Metcalfe looked at me with wide eyes. "Wow!" he said. "Someone's been feeding you on tiger's milk. What's your interest in the safety of Estrenoli?" He looked at me closely, then laughed and snapped his fingers. "Of course, the Contessa has turned chicken."

"Leave her out of it," I said.

Metcalfe smiled shyly, "Ah, you young folk; there's no knowing what you'll get up to next."

"Shut up."

He held up his hands in mock terror. "All right, all right." He laughed suddenly. "You damn' near killed Estrenoli yourself. If you'd have hit him a fraction harder he'd have been a dead man."

"I couldn't hit him harder."

"I wouldn't take any bets on that," said Metcalfe. "He's still unconscious. The quack has wired up his jaw and he won't be able to speak for a month." He poured out another couple of drinks. "All right, I'll see he gets to Rome not hurt any more than he is now."

"I'll want that in writing," I said. "From Estrenoli himself—through the post in a letter from Rome datemarked not later than a week to-day."

Metcalfe was still. "You're pushing it a bit hard, aren't you?" he said softly.

"That's what I want," I said stubbornly.

He looked at me closely. "Someone's been making a man out of you, Hal," he said. "All right; that's the way it'll be." He pushed the drink across the table. "You know," he said musingly, almost to himself, "I wouldn't stay long in Rapallo if I were you. I'd get that keel fixed damn' quick and I'd clear out. Torloni's a bad man to tangle with."

"I'm not tangling with Torloni; I only saw him for the first time to-day."

He nodded. "O.K. If that's the way you're going to play it, that's your business. But look, Hal; you pushed me just now and I played along because Estrenoli is no business of mine and you're by way of being a pal and maybe I don't mind being pushed in this thing. But don't try to push Torloni; he's bad, he'd eat you for breakfast."

"I'm not pushing Torloni," I said. "Just as long as he doesn't push me." I finished the drink and stood up. "I'll see you around."

Metcalfe grinned. "You certainly will. As you said—it's a small town."

He came up on deck to see me off and as I drove back to the yard I wondered greatly about Metcalfe. There had been some plain speaking—but not plain enough—and the whole mystery of Metcalfe's position was deepened. He had as much as said, "Get clear before Torloni chops you," and I couldn't understand his motives—after all, Torloni was *his* man.

It was beyond me.

When I got back to the boatyard work in the shed was continuing as though there had never been a break. There was a sudden glare as a chunk of gold melted and Coertze bent over the mould to pour it.

Francesca came up to me and I said, "It's fixed; you'll hear from Eduardo within the week."

She sighed. "Come and have supper. You haven't eaten yet."

"Thanks," I said and followed her to the caravan.

We worked, my God, how we worked.

The memory of that week remains with me as a dark and shadowed mystery puncutated by bright flashes of colour. We melted and poured gold for sixteen hours a day, until our arms were weary and our eyes sore from the flash of the furnaces. We dropped into our berths at night, asleep before we hit the pillows, and it would seem only a matter of minutes before we were called again to that damned assembly line I had devised.

I grew to hate the sight and the feel of gold, and the smell too—it has a distinctive odour when molten—and I prayed for the time when we would be at sea again with nothing more than a gale and a lee shore to worry about. I would rather have been alone in a small boat in a West Indies hurricane than undergo another week of that torture.

But the work got done. The mass of gold in the mould grew bigger and bigger and the pile of unmelted ingots became smaller. We were doing more than 250 melts a day and I calculated that we would gain half a day on my original ten-day schedule. A twelve-hour gain was not much, but it might mean the difference between victory and defeat.

Metcalfe and Torloni were keeping oddly quiet. We were watched—or rather, the boatyard was watched—and that was all. In spite of the reinforcements that Torloni had pushed into Rapallo, and in spite of the fact that he was personally supervising operations himself, there were no overt moves against us.

I couldn't understand it.

The only cheering aspect of the whole situation was Francesca. She cooked our food and did our housekeeping, received messages and issued instructions to the intelligence service and, although in the pace of work we had little time to be together, there was always something small like a hand's touch or a smile across the room to renew my will to go on.

Five days after I had seen Metcalfe she received a letter which she burned after reading it with a frown of pain on her brow. She came to me and said, "Eduardo is safe in Rome."

"Metcalfe kept his promise," I said.

A brief smile touched her lips. "So will I." She grew serious. "You must see the doctor to-morrow."

"I haven't time," I said impatiently.

"You must make time," she insisted. "You will have to sail *Sanford* very soon; you must be fit."

She brought Coertze into the argument. He said, "She's right. We don't want to depend on Walker, do we?"

That was another worry. Walker was deteriorating rapidly. He was moody and undependable, given to violent tempers and unpredictable fits of sulking. The gold was rotting him slowly but certainly, corrupting him far more surely than any alcohol.

Coertze said, "Man, go to the doctor." He smiled sheepishly. "It's my fault you have a bad back, anyway. I could have shored up that passage better than I did. You go, and I'll see the work doesn't suffer."

That was the first time that Coertze admitted responsibility for anything, and I respected him for it. But he had no sympathy for my scraped knuckles, maintaining that a man should learn how to punch without damaging himself.

So the next day Francesca drove me to see the doctor. After he had hissed and tutted and examined and rebandaged my back, he expressed satisfaction at my progress and said I must see him at the same time the following week. I said I would come, but I knew that by then we would be at sea on our way to Tangier.

When we were again seated in the car Francesca said, "Now we go to the Hotel Levante."

"I've got to get back," I said.

"A drink will do you good," she said. "A few minutes won't hurt."

So we went to the Hotel Levante, wandered into the lounge and ordered drinks. Francesca toyed with her glass and then said hesitatntly, "There's something else—another reason why I brought you here. I want you to meet someone."

"Someone here? Who?"

"My father is upstairs. It is right that you see him."

This was unexpected. "Does he know about us?"

She shook her head. "I told him about the gold and the jewels. He was very angry about that, and I don't know what he is going to do. I did not tell him about you and me."

This looked as though it was going to be a difficult interview. It is not often that a prospective son-in-law has to admit that he is a gold smuggler before he asks for a hand in marriage—a hand that is already married to someone else, to make things worse.

I said, "I would like very much to meet your father."

We finished our drinks and went up to the old man's

room. He was sitting in an arm-chair with a blanket across his knees and he looked up sharply when we appeared. He looked tired and old; his hair was white and his beard no longer bristled, as I had heard it described, but had turned wispy and soft. His eyes were those of a beaten man and had no fight in them.

"This is Mr. Halloran," said Francesca.

I walked across to him. "I'm very glad to meet you, sir."

Something sparked in his eyes. "Are you?" he said, ignoring my outstretched hand. He leaned back in his chair. "So you are the thief who is stealing my country's gold."

I felt my jaw tightening. I said evenly, "Apparently you do not know the laws of your own country, sir."

He raised shaggy white eyebrows. "Oh! Perhaps you can enlighten me, Mr. Halloran."

"This treasure falls under the legal heading of abandoned property," I said. "According to Italian law, whoever first takes possession of it thereafter is the legal owner."

He mused over that. "I dare say you could be right; but, in that case, why all this secrecy?"

I smiled. "A lot of money is involved. Already the vultures are gathering, even with the secrecy we have tried to keep."

His eyes snapped. "I don't think your law is good, young man. This property was not abandoned; it was taken by force of arms from the Germans. It would make a pretty court case indeed."

"The whole value would go in legal expenses, even if we won," I said dryly.

"You have made your point," he said. "But I don't like it, and I don't like my daughter being involved in it."

"Your daughter has been involved in worse things," I said tightly.

"What do you mean by that?" he demanded sharply.

"I mean Estrenoli."

He sighed and leaned back in his chair, the spark that had been in him burned out and he was once more a weary old man. "Yes, I know," he said tiredly. "That was a shameful thing. I ought to have forbidden it, but Francesca . . ."

"I had to do it," she said.

"Well you won't have to worry about him any more," I said. "He'll stay away from you now."

The count perked up. "What happened to him?"

There was a ghost of a smile round Francesca's mouth as she said, "Hal broke his jaw."

"You did? You did?" The Count beckoned. "Come here, young man; sit close to me. You really hit Estrenoli? Why?"

"I didn't like his manners."

He chuckled. "A lot of people don't like the Estrenoli manners, but no one has hit an Estrenoli before. Did you hurt him?"

"A friend tells me that I nearly killed him."

"Ah, a pity," said the Count ambiguously. "But you will have to be careful. He is a powerful man with powerful friends in the Government. You will have to leave Italy quickly."

"I will leave Italy, but not because of Estrenoli. I imagine he is a very frightened man now. He will be no trouble."

The Count said, "Any man who can get the better of an Estrenoli must have my thanks—and my deepest respect."

Francesca came over to me and put her hand on my shoulder. "I also am going to leave Italy," she said. "I am going away with Hal."

The Count looked at her for a long time then dropped his head and stared at the bony hands crossed in his lap. "You must do what you think best, my child," he said in a low voice. "Italy has given you nothing but unhappiness; perhaps to find happiness you must go to another country and live under different laws."

He raised his head. "You will cherish her, Mr. Halloran?"

I nodded, unable to speak.

Francesca went to him, kneeling at his side, and took his hands in hers. "We must do it, Papa; we're in love. Can you give us your blessing?"

He smiled wryly. "How can I give my blessing to something I think is a sin, child? But I think that God is wiser than the churchmen and He will understand. So you have my blessing and you must hope that you have God's blessing too."

She bent her head and her shoulders shook. He looked up at me. "I was against this marriage to Estrenoli, but she did it for me. It is our law here that such a thing cannot be undone."

Francesca dried her eyes. She said, "Papa, we have little time and I must tell you something. Cariaceti—you remember little Cariaceti—will come to you from time to time and give you money. You must . . ."

He broke in. "I do not want such money."

"Papa, listen. The money is not for you. There will be a lot of money and you must take a little for yourself if you need it, but most of it must be given away. Give some to Mario Pradelli for his youngest child who was born spastic;

give some to Pietro Morelli for his son whom he cannot afford to send to university. Give it to those who fought with you in the war; those who were cheated by the Communists just like you were; those who need it."

I said, "My share of the gold is Francesca's, to do with as she likes. That can be added, too."

The Count thought deeply for a long time, then he said musingly, "So something good may come out of this after all. Very well, I will take the money and do as you say."

She said, "Piero Morese will help you—he knows where all your old comrades are. I will not be here; I leave with Hal in a few days."

"No," I said. "You stay. I will come back for you."

"I am coming with you," she declared.

"You're staying here. I won't have you on *Sanford*."

The Count said, "Obey him, Francesca. He knows what he must do, and perhaps he could not do it if you were there."

She was rebellious, but she acquiesced reluctantly. The Count said, "Now you must go, Francesca. I want to talk to your Hal—alone."

"I'll wait downstairs," she said.

The Count watched her go. "I think you are an honourable man, Mr. Halloran. So I was told by Piero Morese when he talked to me on the telephone last night. What are your exact intentions when you take my daughter from Italy?"

"I'm going to marry her," I said. "Just as soon as she can get a divorce."

"You realise that she can never come back to Italy in those circumstances? You know that such a marriage would be regarded here as bigamous?"

"I know—and Francesca knows. You said yourself that she has had nothing but unhappiness in Italy."

"That is true." He sighed. "Francesca's mother died when she was young, before the war. My daughter was brought up in a partisan camp in the middle of a civil war and she has seen both the heroism and degradation of men from an early age. She is not an ordinary woman because of this; some would have been made bitter by her experiences, but she is not bitter. Her heart is big enough to have compassion for all humanity—I would not like to see it broken."

"I love Francesca," I said. "I will not break her heart, not wittingly."

He said, "I understand you are a ship designer and a shipbuilder."

"Not ships—small boats."

"I understand. After I talked to Piero I thought I would see what sort of a man you were, so a friend kindly asked some questions for me. It seems you have a rising reputation in your profession."

I said, "Perhaps in South Africa; I didn't know I was known here."

"There has been some mention of you," he said. "The reason I bring this up is that I am pleased that it is so. This present venture in which you are engaged I discount entirely. I do not think you will succeed—but if you do, such wealth is like the gold of fairies, it will turn to leaves in your hands. It is good to know that you do fine work in the field of your choice."

He pulled the blanket round him. "Now you must go; Francesca will be waiting. I cannot give you more than my good wishes, but those you have wherever you may be."

I took his proffered hand and said impulsively, "Why don't you leave Italy, too, and come with us?"

He smiled and shook his head. "No, I am old and the old do not like change. I cannot leave my country now, but thank you for the thought. Good-bye, Hal, I think you will make my daughter very happy."

I said good-bye and left the room. I didn't see the Count ever again.

II

The time arrived when, incredibly, the keel was cast.

We all stood round the mould and looked at it a little uncertainly. It seemed impossible that all our sweat and labour should have been reduced to this inert mass of dull yellow metal, a mere eight cubic feet shaped in a particular and cunning way.

I said, "That's it. Two more days and *Sanford* will be in the water."

Coertze looked at his watch. "We've got time to do some more work to-day; we can't knock off just because the keel is finished—there's still plenty to do."

So we got on with it. Walker began to strip the furnaces and I directed Coertze and Piero in stripping the glass-fibre cladding from *Sanford* preparatory to removing the lead keel. We were happy that night. The change of work and pace had done us good and we all felt rested.

Francesca reported that everything was quiet on the potential battle-front—Metcalfe was on the Fairmile and Torloni was in his hotel; the watch on the boatyard had not been intensified—in fact, everything was as normal as a thoroughly abnormal situation could be.

The trouble would come, if it had to come at all, when we launched *Sanford*. At the first sign of us getting away the enemy would be forced to make a move. I couldn't understand why they hadn't jumped us before.

The next day was pure joy. We worked as hard as ever and when we had finished *Sanford* was the most expensively built boat in the world. The keel bolts which Coertze had cast into the golden keel slipped smoothly into the holes in the keelson which Harry had prepared long ago in Cape Town, and as we let down the jacks *Sanford* settled comfortably and firmly on to the gold.

Coertze said, "I can't see why you didn't use the existing holes—the ones drilled for the old keel."

"It's the difference in weight distribution," I said. "Gold is half as heavy again as lead and so this keel had to be a different shape from the old one. As it is, I had to juggle with the centre of gravity. With the ballast being more concentrated I think *Sanford* will roll like a tub, but that can't be helped."

I looked at *Sanford*. She was now worth not much short of a million and three-quarter pounds—the most expensive 15-tonner in history. I felt quite proud of her—not many yacht designers could boast of such a design.

When we had supper that night we were all very quiet and relaxed. I said to Francesca, "You'd better get the jewels out to-night—it may be your last chance before the fireworks start."

She smiled. "That will be easy; Piero has cast them into concrete bricks—we are learning the art of disguise from you. They are outside near the new shed that Palmerini is building."

I laughed. "I must see this."

"Come," she said. "I will show you."

We went into the dark night and she flashed a torch on an untidy heap of bricks near the new shed. "There they are; the valuable bricks are spotted with whitewash."

"Not bad," I said. "Not bad at all."

She leaned against me and I put my arms around her. It was not often we had time for this sort of thing, we were missing a lot that normal lovers had. After a moment she said quietly, "When are you coming back?"

"As soon as I've sold the gold," I said. "I'll take the first plane out of Tangier."

"I'll be waiting," she said. "Not here—I'll be in Milan with my father."

She gave me the address which I memorised. I said, "You won't mind leaving Italy?"

"No, not with you."

"I asked your father to come with us, but he wouldn't."

"Not after seventy years," she said. "It's too much to ask an old man."

I said, "I knew that, but I thought I'd make the offer."

We talked for a long time there in the darkness, the small personal things that lovers talk about when they're alone. Then Francesca said that she was tired and was going to bed.

"I'll stay and have another cigarette," I said. "It's pleasant out here."

I watched her melt into the darkness and then I saw the gleam of light as she opened the door of the shed and slipped inside.

A voice whispered from out of the darkness, "Halloran!"

I started, "Who's that?" I flashed my torch about.

"Put out that damned light. It's me—Metcalfe."

I clicked off the torch and stooped to pick up one of the concrete bricks. I couldn't see if it had spots of whitewash on it or not; if it had, then Metcalfe was going to be clobbered by a valuable brick.

A dark silhouette moved closer. "I thought you'd never stop making love to your girl-friend," said Metcalfe.

"What do you want, and how did you get here?"

He chuckled. "I came in from the sea—Torloni's boys are watching the front of the yard."

"I know," I said.

There was surprise in his voice. "Do you, now?" I saw the flash of his teeth. "That doesn't matter, though; it won't make any difference."

"It won't make any difference to what?"

"Hal, boy, you're in trouble," said Metcalfe. "Torloni's going to jump you—to-night. I tried to hold him in, but he's got completely out of hand."

"Whose side are you on?" I demanded.

He chuckled. "Only my own," he said. He changed his tone. "What are you going to do?"

I shrugged. "What can I do except fight?"

"Be damned to that," he said. "You wouldn't have a chance against Torloni's cut-throats. Isn't your boat ready for launching?"

"Not yet. She still needs sheathing and painting."

"What the hell?" he said angrily. "What do you care if you get worm in your planking now? Is the new keel on?"

I wondered how he knew about that. "What if it is?"

"Then get the stick put back and get the boat into the water, and do it now. Get the hell out of here as fast as you can." He thrust something into my hand. "I had your clearance made out. I told you I was a pal of the Port Captain."

I took the paper and said, "Why warn us? I thought Torloni was your boy."

He laughed gently. "Torloni is nobody's boy but his own. He was doing me a favour but he didn't know what was in the wind. I told him I just wanted you watched. I was sorry to hear about the old watchman—that was Torloni's thugs, it wasn't my idea."

I said, "I thought hammering old men wasn't your style."

"Anyway," he said. "Torloni knows the score now. It was that damn' fool Walker who gave it away."

"Walker! How?"

"One of Torloni's men picked his pocket and pinched his cigarette case. It wasn't a bad case, either; it was made of gold and had a nice tasteful inscription on the inside—'*Caro Benito da parte di Adolf—Brennero—1940.*' As soon as Torloni saw that he knew what was up, all right. People have been scouring Italy for that treasure ever since the war, and now Torloni thinks he has it right in his greasy fist."

I damned Walker at length for an incompetent, crazy idiot.

Metcalfe said, "I tried to hold Torloni, but he won't be held any longer. With what's at stake he'd as soon cut my throat as yours— that's why I'm giving you the tip-off."

"When is he going to make his attack?"

"At three in the morning. He's going to move in with all his crowd."

"Any guns?"

Metcalfe's voice was thoughtful. "No, he won't use guns. He wants to do this quietly and he has to get the gold out. That'll take some time and he doesn't want the police breathing down his neck while he's doing it. So there'll be no guns."

That was the only good thing I'd heard since Metcalfe had surprised me. I said, "Where are his men now?"

"As far as I know they're getting some sleep—they don't like being up all night."

"So they're in their usual hotels—all sixteen of them."

Metcalfe whistled. "You seem to know as much about it as I do."

"I've known about it all the time," I said shortly. "We've had them tabbed ever since they moved into Rapallo—before that, too. We had your men spotted in every port in the Mediterranean."

He said slowly, "I wondered about that ever since Dino was beaten up in Monte Carlo. Was that you?"

"Coertze," I said briefly. I gripped the brick which I was still holding. I was going to clobber Metcalfe after all—he played a double game too often and he might be playing one now. I thought we had better keep him where we could watch him.

He laughed. "Yes, of course; that's just his mark."

I lifted the brick slowly. "How did you cotton on to us?" I asked. "It must have been in Tangier, but what gave the game away?"

There was no answer.

I said, "What was it, Metcalfe?" and raised the brick.

There was silence.

"Metcalfe?" I said uncertainly, and switched on my torch. He had gone and I heard a faint splashing from the sea and the squeak of an rowlock. I ought to have known better than to think I could outwit Metcalfe; he was too wise a bird for me.

III

As I went back to the shed I looked at my watch; it was ten o'clock—five hours to go before Torloni's assault. Could we replace the mast and all the standing rigging in time? I very much doubted it. If we turned on the floodlights outside the shed, then Torloni's watchers would know that something unusual was under way and he would move in immediately. If we worked in the dark it would be hell's own job—I had never heard of a fifty-five foot mast being stepped in total darkness and I doubted if it could be done.

It looked very much as though we would have to stay and fight.

I went in and woke Coertze. He was drowsy but he woke up fast enough when I told him what was happening. I omitted to mention Walker's part in the mess—I still needed Walker and I knew that if I told Coertze about it I would have a corpse and a murderer on my hands, and this was no time for internal dissension.

Coertze said suspiciously, "What the hell is Metcalfe's game?"

"I don't know and I care less. The point is that he's given us the tip-off and if we don't use it we're fools. He must have fallen out with Torloni."

"*Reg,*" said Coertze and swung himself out of his berth. "Let's get cracking."

"Wait a minute," I said. "What about the mast?" I told him my estimate of the chance of replacing the mast in darkness.

He rubbed his chin and the bristles crackled in the silence. "I reckon we should take a chance and turn the lights on," he said at last. "That is, after we've made our preparations for Torloni. We know he's going to attack and whether he does it sooner or later doesn't matter as long as we're ready for him."

This was the man of action—the military commander—speaking. His reasoning was good so I left him to it. He roused Piero and they went into a huddle while Walker and I began to clear the shed and to load up *Sanford.* Francesca heard the noise and got up to see what was going on and was drawn into Coertze's council of war.

Presently Piero slipped out of the shed and Coertze called me over. "You might as well know what's going to happen," he said.

He had a map of Rapallo spread out, one of the give-aways issued by the Tourist Office, and as he spoke he pointed to the salient features on the map. It was a good plan that he described and like all good plans it was simple.

I think that if Coertze had not been taken prisoner at Tobruk he would have been commissioned as an officer sooner or later. He had a natural grasp of strategy and his plan was the classic military design of cencentration to smash the enemy in detail before they could concentrate.

He said, "This is the holiday season and the hotels are full. Torloni couldn't get all his men into the same hotel, so they're spread around the town—four men here, six here, three here and the rest with Torloni himself." As he spoke his stubby forefinger pointed to places on the map.

"We can call up twenty-five men and I'm keeping ten men here at the yard. There are four of Torloni's men outside the yard right now, watching us, and we're going to jump them in a few minutes—ten men should clean them up easily. That means that when we turn on the lights there'll be no one to warn Torloni about it."

"That seems a good idea," I said.

"That leaves us fifteen men we can use outside the yard as a mobile force. We have two men outside each hotel excepting this one, here, where we have nine. There are four of Torloni's men staying here and when they come out they'll get clobbered. That ought to be easy, too."

"You'll have already cut his force by half," I said.

"That's right. Now, there'll be Torloni and eight men moving in on the yard. He'll expect to have sixteen, but he won't get them. This may make him nervous, but I think not. He'll think that there'll only be four men and a girl here and he'll reckon he can take us easily. But we'll have fourteen men in the yard—counting us—and I'll bring in another fifteen behind him as soon as he starts anything."

He looked up. "How's that, ay?"

"It's great," I said. "But you'll have to tell the Italians to move in fast. We want to nail those bastards quick before they can start shooting. Metcalfe said they wouldn't shoot, but they might if they see they're on the losing end."

"They'll be quick," he promised. "Piero's on the blower now, giving instructions. The orders are to clean up the four watchers here at eleven o'clock." He looked at his watch. "That's in five minutes. Let's go and see the fun."

Francesca said, "I don't see how anything can go wrong."

Neither could I—but it did!

We were leaving the shed when I noticed Walker tagging on behind. He had been keeping in the background, trying to remain inconspicuous. I let the others go and caught his arm. "You stay here," I said. "If you move out of this shed I swear I'll kill you."

His face went white. "Why?"

"So you had your wallet stolen," I said. "You damn' fool, why did you have to carry that cigarette case?"

He tried to bluff his way out of it. "Wh . . . what cigarette case?"

"Don't lie to me. You know what cigarette case. Now stay here and don't move out. I don't want you underfoot—I don't want to have to keep an eye on you all the time in case you make any more damn' silly mistakes." I took him by the shirt. "If you don't stay in here I'll tell Coertze just why Torloni is attacking to-night—and Coertze will dismember you limb from limb."

His lower lip started to tremble. "Oh, don't tell Coertze," he whispered. "Don't tell him."

I let him go. "O.K. But don't move out of this shed."

I followed the others up to Palmerini's office. Coertze said, "It's all set."

I said to Piero, "You'd better get Palmerini down here; we'll need his help in rigging the mast."

"I have telephoned him," said Piero. "He will be coming at eleven-fifteen—after we have finished our work here." He nodded towards the main gate.

"Fine," I said. "Do you think we shall see anything of what is happening?"

"A little. One of Torloni's men is not troubling to hide himself; he is under the street lamp opposite the main gate."

We went up to the gate, moving quietly so as not to alarm the watchers. The gate was of wood, old, unpainted and warped by the sun; there were plenty of cracks through which we could see. I knelt down and through one of the cracks saw a man on the other side of the road, illuminated by the street lamp. He was standing there, idly smoking a cigarette, with one hand in his trouser-pocket. I could hear the faint click as he jingled money or keys.

Coertze whispered, "Any time now."

Nothing happened for a while. There was no sound to be heard except for the sudden harsh cry of an occasional seabird. Piero said in a low voice. "Two have been taken."

"How do you know?"

There was laughter in his voice. "The birds—they tell me."

I suddenly realised what had been nagging at my mind. Seagulls sleep at night and they don't cry.

There was a faint sound of singing which grew louder, and presently three men came down the street bellowing vociferously. They had evidently been drinking because they wavered and staggered and one of them had to be helped by the others. The man under the lamp trod on the butt of his cigarette and moved back to the wall to let them pass. One of them waved a bottle in the air and shouted, "Have a drink, brother; have a drink on my first-born."

Torloni's man shook his head but they pressed round him clamouring in drunken voices for him to drink. Suddenly the bottle came down sharply and I heard the thud even from across the street.

"God," I said. "I hope they haven't killed him."

Piero said, "It will be all right; they know the thickness of a man's skull."

The drunken men were suddenly miraculously sober and came across the street at a run carrying the limp figure of Torloni's man. Simultaneously others appeared from the left and the right, also bearing unconscious bodies. A car came up the street and swerved through the gateway.

"That's four," said Coertze with satisfaction. "Take them into the shed."

"No," I said. "Put them in that half-finished shed." I didn't want them to get a glimpse of anything that might do us damage later. "Tie them up and gag them; let two men watch them."

Piero issued orders in rapid Italian and the men were carried away. We were surrounded by a group of Italians babbling of how easy it all was until Piero shouted for silence. "Are you veterans or are you green recruits?" he bawled. "By God, if the Count could see you now he'd have you all shot."

There was an abashed silence at this, and Piero said, "Keep a watch outside. Giuseppi, go to the office and stay with the telephone; if it rings, call me. You others, watch and keep quiet."

A car hooted outside the gate and I started nervously. Piero took a quick look outside. "It is all right; it is Palmerini. Let him in."

Palmerini's little Fiat came through the gateway and disgorged Palmerini and his three sons in a welter of arms and legs. He came up to me and said, "I am told you are in a hurry to get your boat ready for sea. That will be extra for the overtime, you understand."

I grinned. Palmerini was running true to form. "How long will it take?"

"With the lights—four hours, if you help, too."

That would be three-fifteen—just too late. We would probably have to fight, after all. I said, "We may be interrupted, Signor Palmerini."

"That is all right, but any damage must be paid for," he answered.

Evidently he knew the score, so I said, "You will be amply recompensed. Shall we begin?"

He turned and began to berate his sons. "What are you waiting for, you lazy oafs; didn't you hear the signor? The good God should be ashamed for giving me sons so strong in the arm but weak in the head." He chased them down to the shed and I began to feel happier about everything.

As the lights sprang up at the seaward end of the shed

Francesca looked at the gate and said thoughtfully, "If I was Torloni and I wanted to come in here quickly I would drive right through the gate in a car."

"You mean ram it?"

"Yes, the gate is very weak."

Coertze said jovially, "*Reg*, we can soon stop that. We've captured one of his cars; I'll park it across the gateway behind the gate. If he tries that trick he'll run into something heavier than he bargains for."

"I'll leave you to it, then," I said. "I've got to help Palmerini." I ran down to the shed and heard the car revving up behind me.

Palmerini met me at the door of the shed. He was outraged. "Signor, you cannot put this boat into the water. There is no paint, no copper, nothing on the bottom. She will be destroyed in our Mediterranean water—the worms will eat her up entirely."

I said, "We have no time; she must go into the water as she is."

His professional ethics were rubber raw. "I do not know whether I should permit it," he grumbled. "No boat has ever left my yard in such a condition. If anyone hears of it they will say, 'Palmerini is an old fool; Palmerini is losing his mind—he is getting senile in his old age.'"

In my impatience to get on with the job I suspected he wasn't far off the truth. I said, "No one will know, Signor Palmerini. I will tell no one."

We walked across to *Sanford*. Palmerini was still grumbling under his breath about the iniquity of leaving a ship's bottom unprotected against the small beasts of the sea. He looked at the keel and rapped it with his knuckles. "And this, signor. Whoever heard of a brass keel?"

"I told you I was experimenting," I said.

He cocked his head on one side and his walnut face looked at me impishly. "Ah, signor, never has there been such a yacht as this in the Mediterranean. Not even the famous *Argo* was like this boat, and not even the Golden Fleece was so valuable." He laughed. "I'll see if my lazy sons are getting things ready."

He went off into the lighted area in front of the shed, cackling like a maniac. I suppose no one could do anything in his yard without his knowing exactly what was going on. He was a great leg-puller, this Palmerini.

I called him back, and said, "Signor Palmerini, if all goes

well I will come back and buy your boatyard if you are willing to sell. I will give you a good price."

He chuckled. "Do you think I would sell my yard to a man who would send a boat out without paint on her bottom? I was teasing you, my boy, because you always look so serious."

I smiled. "Very well, but there is a lead keel I have no use for. I'm sure you can use it." At the current price of lead the old keel was worth nearly fifteen hundred pounds.

He nodded judiciously. "I can use it," he said. "It will just about pay for to-night's overtime." He cackled again and went off to crack the whip over his sons.

Walker was still sullen and pale and when I began to drive him he became even more sulky, but I ignored that and drove him all the more in my efforts to get *Sanford* ready for sea. Presently we were joined by Coertze and Francesca and the work went more quickly.

Francesca said, "I've left Piero in charge up there. He knows what to do; besides, he knows nothing about boats."

"Neither do you," I said.

"No, but I can learn."

I said, "I think you should leave now. It might get a bit dangerous round here before long."

"No," she said stubbornly, "I'm staying."

"You're going."

She faced me. "And just how will you make me go?"

She had me there and she knew it. I hesitated, and she said, "Not only am I staying, but I'm coming with you in *Sanford*.'

"We'll see about that later," I said. "At the moment I've no time to argue."

We pulled *Sanford* out of the shed and one of Palmerini's sons ran the little crane alongside. He picked up the mast and hoisted it high above the boat, gently lowering it between the mast partners. I was below, making sure that the heel of the mast was correctly bedded on the butt plate. Old Palmerini came below and said, "I'll see to the wedges. If you are in the hurry you say you are, you had better see that your engine is fit to run."

So I went aft and had a look at the engine. When *Sanford* had been taken from the water I had checked the engine twice a week, turning her over a few revolutions to circulate the oil. Now, she started immediately, running sweetly, and I knew with satisfaction that once we were in the water we could get away at a rate of knots.

I checked the fuel tanks and the water tanks and then went on deck to help the Palmerini boys with the standing rigging. After we had been working for some time, Francesca brought us coffee. I accepted it with thanks, and she said quietly, "It's getting late."

I looked at my watch; it was two o'clock. "My God!" I said. "Only an hour before the deadline. Heard anything from Piero?"

She shook her head. "How long will it be before you are finished?" she asked, looking round the deck.

"It looks worse than it is," I said. "I reckon we'll be nearly two hours, though."

"Then we fight," she said with finality.

"It looks like it." I thought of Coertze's plan. "It shouldn't come to much, though."

"I'll stay with Piero," she said. "I'll let you know if anything happens."

I watched her go, then went to Walker. "Never mind the running rigging," I said. "We'll fix that at sea. Just reeve the halyards through the sheaves and lash them down. We haven't much time now."

If we worked hard before, we worked harder then—but it was no use. Francesca came running down from the office. "Hal, Hal, Piero wants you."

I dropped everything and ran up the yard, calling for Coertze as I went. Piero was talking on the telephone when I arrived. After a minute he hung up and said, "It's started."

Coertze sat on at the desk upon which was spread the map. "Who was that?"

Piero laid his finger on the map. "These men. We have two men following."

"Not the four we're tackling straight away?" I asked.

"No, I haven't heard of them." He crossed to the window and spoke a few words to a man outside. I looked at my watch—it was half past two.

We sat in silence and listened to the minutes tick away. The atmosphere was oppressive and reminded me of the time during the war when we expected a German attack but didn't know just when or where it was going to come.

Suddenly the telephone rang and we all started.

Piero picked it up and as he listened his lips tightened. He put the telephone down and said, "Torloni has got more men. They are gathering in the Piazza Cavour—there are two lorry loads."

"Where the hell did *they* come from?" I demanded.

"From Spezia; he has called in another gang."

My brain went into high gear. Why had Torloni done that? He didn't need so many men against four of us—unless he knew of our partisan allies—and it was quite evident that he did. He was going to overrun us by force of numbers.

"How many extra men?" asked Coertze.

Piero shrugged. "At least thirty, I was told."

Coertze cursed. His plan was falling to pieces—the enemy was concentrating and our own forces were divided.

I said to Piero, "Can you get in touch with your men?"

He nodded. "One watches—the other is near a telephone."

I looked at Coertze. "You'd better bring them in."

He shook his head violently. "No, the plan is still good. We can still engage them here and attack them in the rear."

"How many men have we got altogether?"

Coertze said, "Twenty-five Italians and the four of us."

"And they've got forty-three, at least. Those are bad odds."

Francesca said to Piero. "The men we have are those who can fight. There are others who cannot fight but who can watch. It is a pity that the fighters have to be watchers, too. Why not get some of the old men to do the watching so that you can collect the fighters together?"

Piero's hand went to the phone but stopped as Coertze abruptly said, "No!" He leaned back in his chair. "It's a good idea, but it's too late. We can't start changing plans now. And I want that phone free—I want to know what is happening to our mobile force."

We waited while the leaden minutes dragged by. Coertze suddenly said, "Where's Walker?"

"Working on the boat," I said. "He's of more use down there."

Coertze snorted. "That's God's truth. He'll be no use in a brawl."

The telephone shrilled and Piero scooped it up in one quick movement. He listened intently, then began to give quick instructions. I looked at Coertze and said, "Four down."

". . . and thirty-nine to go," he finished glumly.

Piero put down the phone. "That was the mobile force—they are going to the Piazza Cavour."

The phone rang again under his hand and he picked it up. I said to Francesca, "Go down to the boat and tell Walker to work like hell. You'd better stay down there, too."

As she left the office, Piero said, "Torloni has left the Piazza

Cavour—two cars and two trucks. We had only two men there and they have already lost one truck. The other truck and the cars are coming straight here."

Coertze thumped the table. "Dammit, where did that other truck go?"

I said sardonically, "I wouldn't worry about it. Things can't help but get better from now on; they can't get any worse, and we've nowhere to go but up."

I left the office and stood in the darkness. Giuseppi said, "What is happening, signor?"

"Torloni and his men will be here within minutes. Tell the others to be prepared."

After a few moments Coertze joined me. "The telephone line's been cut," he said.

"That tops it," I said. "Now we don't know what's going on at all."

"I hope our friends outside use their brains and concentrate into one bunch; if they don't, were sunk," he said grimly.

Piero joined us. "Will Palmerini's sons fight?" I asked.

"Yes, if they are attacked."

"You'd better go down and tell the old man to lie low. I wouldn't want him to get hurt."

Piero went away and Coertze settled down to watch. The street was empty and there was no sound. We waited a long time and nothing happened at all. I thought that perhaps Torloni was disconcerted by finding his watchmen missing—that might put him off his stroke. And if he had a roll-call and discovered a total of eight men missing it was bound to make him uneasy.

I looked at my watch—three-fifteen. If Torloni would only hold off we might get the boat launched and away and the men dispersed. I prayed he would hold off at least another half-hour.

He didn't.

Coertze suddenly said, "Something's coming."

I heard an engine changing gear and the noise was suddenly loud. Headlights flashed from the left, approaching rapidly, and the engine roared. I saw it was a lorry being driven fast, and when it was abreast of the yard, it swerved and made for the gate.

I blessed Francesca's intuition and shouted in Italian, "To the gates!"

The lorry smashed into the gates and there was a loud cracking and snapping of wood, overlaid by the crash as the lorry hit the car amidships and came to a jolting halt. We didn't wait for Torloni's men to recover but piled in immediately. I scrambled over the ruined car and got on to the bonnet of the lorry, whirling round to the passenger side. The man in the passenger seat was shaking his head groggily; he had smashed it against the windscreen, unready for such a fierce impact. I hit him with my fist and he slumped down to the floor of the cab.

The driver was frantically trying to restart his stalled engine and I saw Coertze haul him out bodily and toss him away into the darkness. Then things got confused. Someone from the back of the lorry booted me on the head and I slipped from the running-board conscious of a wave of our men going in to the attack. When I had recovered my wits it was all over.

Coertze dragged me from under the lorry and said, "Are you all right?"

I rubbed my sore head. "I'm O.K. What happened?"

"They didn't know what hit them—or they didn't know what they hit. The smash shook them up too much to be of any use; we drove them from the lorry and they ran for it."

"How many of them were there?"

"They were jammed in the back of the lorry like sardines. I suppose they thought they could smash in the gates, drive into the yard and get out in comfort. They didn't get the chance." He looked at the gateway. "They won't be coming that way again."

The gateway, from being our weakest point had become our strongest. The tangled mess of the lorry and the car completely blocked the entrance, making it impassable.

Piero came up and said, "We have three prisoners."

"Tie them up and stick them with the others," I said. One commodity which is never in short supply in a boatyard is rope. Torloni was now missing eleven men—a quarter of his force. Perhaps that would make him think twice before attacking again.

I said to Coertze, "Are you sure they can't attack us from the sides?"

"Positive. We're blocked in with buildings on both sides. He has to make a frontal attack. But, hell, I wish I knew where that other lorry went."

The telephone began to ring shrilly.

I said, "I thought you said the wire had been cut."

"Piero said it had."

We ran to the office and Coertze grabbed the phone. He listened for a second, then said, "It's Torloni!"

"I'll speak to him," I said, and took the phone. I held my

hand over the mouthpiece. "I've got an idea—get old Palmerini up here." Then I said into the phone, "What do you want?"

"Is that Halloran?" The English was good, if strongly tinged with an American accent.

"Yes."

"Halloran, why don't you be reasonable? You know you haven't a chance."

I said, "This phone call of yours is proof that we *have* a chance. You wouldn't be speaking to me if you thought you could get what you want otherwise. Now, if you have a proposition, make it; if you haven't, shut up."

His voice was softly ugly. "You'll be sorry you spoke to me like that. Oh, I know all about the Estrenoli woman's old soldiers, but you haven't got enough of them. Now if you cut me in for half I'll be friendly."

"Go to hell!"

"All right," he said. "I'll crush you and I'll like doing it."

"Make one more attack and the police will be here." I might as well try to pull a bluff.

He thought that one over, then said silkily, "And how will you call them with no telephone?"

"I've made my arrangements," I said. "You've already run into some of them." I rubbed it in. "A lot of your men are mysteriously missing, aren't they?"

I could almost hear his brain click to a decision. "You won't send for the police," he said with finality. "You want the police as little as I do. Halloran, I did you a favour once; I got rid of Estrenoli, didn't I? You could return the favour."

"The favour was for Metcalfe, not me," I said, and hung up on him. He wouldn't like that.

Coertze said, "What did he want?"

"A half-share—or so he said."

"I'll see him in hell first," he said bluntly.

"Where's Palmerini?"

"Coming up. I sent Giuseppi for him."

Just then Palmerini came into the office. I said, "First, how's the boat getting on?"

"Give me fifteen minutes—just fifteen minutes, that's all."

"I may not be able to," I said. "You've got some portable floodlights you use for working at night. Take two men and bring them up here quickly."

I turned to Coertze. "We want to be able to see what's happening. They'll have to come over the wall this time, and once they're over it won't be easy for them to get back. That means that the next attack will be final—make or break. Now here's what we do."

I outlined what I wanted to do with the lights and Coertze nodded appreciatively. It took a mere five minutes to set them up and we used the Fiat and a truck to give added light by their headlamps. We placed the men and settled down to wait for the impending attack.

It wasn't long in coming. There were odd scraping noises from the wall and Coertze said, "They're coming over."

"Wait," I breathed.

There were several thumps which could only be made by men dropping heavily to the ground. I yelled, "*Luce!*" and the lights blazed out.

It was like a frozen tableau. Several of the enemy were on our side of the wall, squinting forward at the light pouring on to them. Several others were caught lowering themselves, their heads turned to see what was happening.

What they must have seen cannot have been reassuring—a blaze of blinding light behind which was impenetrable darkness heavy with menace, while they themselves were in the open and easily spotted—not a very comfortable thought for men supposedly making a surprise attack.

They hesitated uncertainly and in that moment we hit them on both flanks simultaneously, Piero leading from the right and Coertze from the left. I stayed with a small reserve of three men, ready to jump in if either flank party had bitten off more than it could chew.

I saw upraised clubs and the flash of knives and three of Torloni's men went down in the first ten seconds. We had caught them off balance and the flank attacks quickly rolled them up into the centre and there was a confused mob of shouting, fighting men. But more of the enemy were coming over the wall fast, and I was just going to move my little group into battle when I heard more shouting.

It came from *behind* me.

"Come on," I yelled and ran down the yard towards *Sanford*. Now we knew what had happened to that other lorry load of men. They had come in from the seaward side and Torloni was attacking us front and rear.

Sanford was besieged. A boat was drawn up on the hard and another boat full of men was just landing. There was a fight going on round *Sanford* with men trying to climb up on to the deck and our working party valiantly trying to drive them off. I saw the small figure of old Palmerini; he had a rope with a block on the end of it which he whirled round his head like a medieval ball and chain. He whirled it once again and the block caught an attacker under the jaw and he toppled from the ladder he was climbing and fell senseless to the ground.

Palmerini's sons were battling desperately and I saw one go down. Then I saw Francesca wielding a boat-hook like a spear. She drove it at a boarder and the spike penetrated his thigh. He screamed shrilly and fell away, the boat-hook still sticking out of his leg. I saw the look of horror on Francesca's face and then drove home my little attack.

It was futile. We managed to relieve the beleaguered garrison on *Sanford,* but then we were outnumbered three to one and had to retreat up the yard. The attackers did not press us; they were so exultant at the capture of *Sanford* that they stayed with her and didn't follow us. We were lucky in their stupidity.

I looked around to see what was happening at the top of the yard. Coertze's party was closer than I had hoped—he had been driven back, too, but he was not under attack and I wondered why. If both enemy groups now made a concerted effort we were lost.

I said to Francesca, "Duck under those sacks and stay quiet—you may get away with it." Then I ran over to Coertze "What's happening?"

He grinned and wiped some blood from his cheek. "Our outside boys concentrated and hit Torloni hard on the other side of the wall, all fifteen of them. He can't retreat now—anyone who tries to go back over the wall gets clobbered. I'm just getting my breath back before I hit 'em again."

I said, "They've got *Sanford*. They came in from the sea—we're boxed in, too."

His chest heaved. "All right; we'll hit 'em down there."

I looked up the yard. "No," I said. "Look, there's Torloni."

We could see him under the wall, yelling at his men, whipping them up for another attack. I said, "We attack up the yard—all of us—and we hope that the crowd at the back of us stay put for the time we need. We're going to snatch Torloni himself. Where's Piero?"

"I am here."

"Good! Tell your boys to attack when I give the signal. You stay with Coertze and me, and the three of us will make for Torloni."

I turned to find Francesca at my elbow. "I thought I told you to duck out of sight."

She shook her head stubbornly. Old Palmerini was behind her, so I said, "See that she stays here, old friend."

He nodded and put his arm round her. I said to Coertze, "Remember, we want Torloni—we don't stop for anything else."

Then we attacked up the yard. The three of us, Coertze, Piero and I, made a flying wedge, evading anyone who tried to stop us. We didn't fight, we just ran. Coertze had grasped the idea and was running as though he was on a rugby field making an effort for the final try.

The goal line was Torloni and we were on to him before he properly realised what was happening. He snarled and blue steel showed in his hand.

"Spread out!" I yelled, and we separated, coming at him from three sides. The gun in his hand flamed and Coertze staggered; then Piero and I jumped him. I raised my arm and hit him hard with the edge of my hand; I felt his collar-bone break and he screamed and dropped the pistol.

With Torloni's scream a curious hush came over the yard. There was an uncertainty in his men as they looked back to see what was happening. I picked up the gun and held it to Torloni's head. "Call off your dogs or I'll blow your brains out," I said harshly.

I was as close to murder then as I have ever been. Torloni saw the look in my eyes and whitened. "Stop," he croaked.

"Louder," ordered Piero and squeezed his shoulder.

He screamed again, then he shouted, "Stop fighting—stop fighting. Torloni says so."

His men were hirelings—they fought for pay and if the boss was captured they wouldn't get paid. There is not much loyalty among mercenaries. There was an uncertain shuffling and a melting away of figures into the darkness.

Coertze was sitting on the ground, his hand to his shoulder. Blood was oozing between his fingers. He took his hand away and looked at it with stupefied amazement. "The bastard shot me," he said blankly.

I went over to him. "Are you all right?"

He held his shoulder again and got to his feet. "I'm O.K." He looked at Torloni sourly. "I've got a bone to pick with you."

"Later," I said. "Let's deal with the crowd at the bottom of the yard."

We were being reinforced rapidly by men climbing over the wall. This was our mobile force which had taken Torloni's men in the rear and had whipped them. In a compact mass we marched down the yard towards *Sanford,* Torloni being frog-marched in front.

As we came near *Sanford* I poked the pistol muzzle into Torloni's fleshy neck. "Tell them," I commanded.

He shouted, "Leave the boat. Go away. Torloni says this."

The men around *Sanford* looked at us expressionlessly and made no move. Piero squeezed Torloni's shoulder again. "Aaah. Leave the boat, I tell you," he yelled.

They raised their eyes to the crowd behind us, realised they were outnumbered, and slowly began to drift towards the hard where their boats were drawn up. Piero said quietly, "These are the men from La Spezia. That man in the blue jersey is their leader, Morlaix; he is a Frenchman from Marseilles." He looked speculatively at their boats. "You may have trouble with him yet. He does not care if Torloni lives or dies."

I watched Morlaix's crowd push their boats into the water. "We'll cross that bridge when we come to it," I said. "We've got to get out of here. Somebody might have notified the police about this brawl—we made enough noise, and there was a gunshot. Did we have many casualties?"

"I don't know; I will find out."

Palmerini came pushing through the crowd with Francesca at his side. "The boat is not harmed," he said. "We can put her into the water at any time."

"Thanks," I said. I looked at Francesca and made a quick decision. "Still want to come?"

"Yes, I'm coming."

"O.K. You won't have time to pack, though. We're leaving within the hour."

She smiled. "I have a small suitcase already packed. It has been ready for a week."

Coertze was standing guard over Torloni. "What do we do with this one?" he asked.

I said, "We take him with us a little way. We may need him yet. Francesca, Kobus was shot; will you strap him up?"

"Oh, I didn't know," she said. "Where is the wound?"

"In the shoulder," said Coertze absently. He was watching Walker on the deck of *Sanford*. "Where was that *kêrel* when the trouble started?"

"I don't know," I said. "I never saw him from start to finish."

IV

We put *Sanford* into the water very easily; there were plenty of willing hands. I felt better with a living, moving deck under my feet than I had for a long time. Before I went aboard for the last time I took Piero on one side.

"Tell the Count I've taken Francesca away," I said. "I think it's better this way—Torloni might look for revenge. You men can look after yourselves, but I wouldn't like to leave her here."

"That is the best thing," he said.

"If Torloni wants to start any more funny tricks you know what to do now. Don't go for his men—go for Torloni. He cracks easily under direct pressure. I'll make it clear to him that if he starts any of his nonsense he'll wind up floating somewhere in the bay. What did you find out about casualties?"

"Nothing serious," said Piero. "One broken arm, three stab wounds, three or four concussions."

"I'm glad no one was killed," I said. "I wouldn't have liked that. I think Francesca would like to speak to you, so I'll leave you to it."

We shook hands warmly and I went aboard. Piero was a fine man—a good man to have beside you in a fight.

He and Francesca talked together for a while and then she came on board. She was crying a little and I put my arm about her to comfort her. It's not very pleasant to leave one's native land at the best of times, and leaving in these circumstances the unpleasantness was doubled. I sat in the cockpit with my hand on the tiller and Walker started the engine. As soon as I heard it throb I threw it into gear and we moved away slowly.

For a long time we could see the little patch of light in front of the shed speckled with the waving Italians. They waved although they could not see us in the darkness and I felt sad at leaving them. "We'll come back sometime," I said to Francesca.

"No," she said quietly. "We'll never be back."

V

We pressed on into the darkness at a steady six knots making our way due south to clear the Portovento headland. I looked up at the mast dimly outlined against the stars and wondered how long it would take to fix the running rigging. The deck was a mess, making nonsense of the term "ship-shape," but we couldn't do anything about that until it was light. Walker was below and Coertze was on the foredeck keeping guard on Torloni. Francesca and I conversed in low tones in the cockpit, talking of when we would be able to get married.

Coertze called out suddenly, "When are we going to get rid of this garbage? He wants to know. He thinks we're going to put him over the side and he says he can't swim."

"We'll slip inshore close to Portovento," I said. "We'll put him ashore in the dinghy."

Coertze grumbled something about it being better to get rid of Torloni there and then, and relapsed into silence. Francesca said, "Is there something wrong with the engine? It seems to be making a strange noise."

I listened and there was a strange noise—but it wasn't *our* engine. I throttled back and heard the puttering of an outboard motor quite close to starboard.

"Get below quickly," I said, and called to Coertze in a low voice, "We've got visitors."

He came aft swiftly. I pointed to starboard and, in the faint light of the newly risen moon, we could see the white feather of a bow wave coming closer. A voice came across the water. "Monsieur Englishman, can you hear me?"

"It's Morlaix," I said, and raised my voice. "Yes, I can hear you."

"We are coming aboard," he shouted. "It is useless to resist."

"You stay clear," I called. "Haven't you had enough?"

Coertze got up with a grunt and went forward. I pulled Torloni's gun from my pocket and cocked it.

"There are only four of you," shouted Morlaix. "And many more of us."

The bow wave of his boat was suddenly much closer and I could see the boat more clearly. It was full of men. Then it was alongside and, as it came close enough to bump gunwales, Morlaix jumped to the deck of *Sanford*. He was only four feet away from me so I shot him in the leg and he gave a shout and fell overboard.

Simultaneously Coertze rose, lifting in one hand the struggling figure of Torloni. "Take this rubbish," he shouted and hurled Torloni at the rush of men coming on deck. Torloni wailed and the flying body bowled them over and they fell back into their boat.

I took advantage of the confusion by suddenly bearing to port and the gap between the boats widened rapidly. Their boat seemed to be out of control—I imagine that the steersman had been knocked down.

They didn't bother us again. We could hear them shouting in the distance as they fished Morlaix from the water, but they made no further attack. They had no stomach for guns.

Our wake broadened in the moonlight as we headed for the open sea. We had a deadline to meet in Tangier and time was short.

VIII. CALM AND STORM

We had fair winds at first and *Sanford* made good time. As I had suspected, the greater concentration of weight in the keel made her crotchety. In a following sea she rolled abominably, going through a complete cycle in two minutes. With the wind on the quarter, usually *Sanford's* best point of sailing, every leeward roll was followed by a lurch in the opposite direction and her mast described wide arcs against the sky.

There was nothing to be done about it so it had to be suffered. The only cure was to have the ballast spread out more and that was the one thing we couldn't do. The violent motion affected Coertze most of all; he wasn't a good sailor at the best of times and the wound in his shoulder didn't help.

With the coming of dawn after that momentous and violent night we lay hove-to just out of sight of land and set to work on the running rigging. It didn't take long—Palmerini had done more in that direction than I'd expected—and soon we were on our way under sail. It was then that the crankiness of *Sanford* made itself evident, and I experimented for a while to see what I could do, but the cure was beyond me so I stopped wasting time and we pressed on.

We soon fell into our normal watchkeeping routine, modified by the presence of Francesca, who took over the cooking from Coertze. During small boat voyages one sees very little of the other members of the crew apart from the times when the watch is changed, but Walker was keeping more to himself than ever. Sometimes I caught him watching me and he would start and roll his eyes like a frightened horse and look away quickly. He was obviously terrified that I would tell Coertze about the cigarette case. I had no such intention—I needed Walker to help run *Sanford*—but I didn't tell him so. Let him sweat, I thought callously.

Coertze's shoulder was not so bad; it was a clean flesh wound and Francesca kept it well tended. I insisted that he sleep in the quarter berth where the motion was least violent, and this led to a general post. I moved to the port pilot berth in the main cabin while Francesca had the starboard pilot berth. She rigged up a sailcloth curtain in front of it to give herself a modicum of privacy.

This meant that Walker was banished to the forecastle to sleep on the hitherto unused pipe berth. This was intended for a guest in port and not for use at sea; it was uncomfortable and right in the bows where the motion is most felt. Serve him damn' well right, I though uncharitably. But it meant that we saw even less of him.

We made good time for the first five days, logging over a hundred miles a day crossing the Ligurian Sea. Every day I shot the sun and contentedly admired the course line on the chart as it stretched even farther towards the Balearics. I derived great pleasure from teaching Francesca how to handle *Sanford*; she was an apt pupil and made no more than the usual beginner's mistakes.

I observed with some amusement that Coertze seemed to have lost his antipathy towards her. He was a changed man, not as prickly as before. The gold was safe under his feet and I think the fight in the boatyard had worked some of the violence out of him. At any rate, he and Francesca got on well together at last, and had long conversations about South Africa.

Once she asked him what he was going to do with his share of the spoil. He smiled. "I'm going to buy a *plaas*," he said complacently.

"A what?"

"A farm," I translated. "All Afrikaners are farmers at heart; they even call themselves farmers—boers—at least they used to."

I think that those first five days after leaving Italy were the best sea days of the whole voyage. We never had better days before and we certainly didn't have any afterwards.

On the evening of the fifth day the wind dropped and the next day it kept fluctuating as though it didn't know what to do next. The strength varied between force three and dead calm and we had a lot of sail work to do. That day we only logged seventy miles.

At dawn the next day there was a dead calm. The sea was slick and oily and coming in long even swells. Our tempers tended to fray during the afternoon when there was nothing to do but watch the mast making lazy circles against the sky, while the precious hours passed and we made no way towards Tangier. I got tired of hearing the squeak of the boom in the gooseneck so I put up the crutch and we lashed down the boom. Then I went below to do some figuring at the chart table.

We had logged twenty miles, noon to noon, and at that rate we would reach Tangier about three months too late. I checked the fuel tank and found we had fifteen gallons left—that would take us 150 miles in thirty hours at our most economical speed. It would be better than sitting still and listening to the halyards slatting against the mast, so I started the engine and we were on our way again.

I chafed at the use of fuel—it was something we might need in an emergency—but this *was* an emergency, anyway, so I might as well use it; it was six of one and half a dozen of the other. We ploughed through the still sea at a steady five knots and I laid a course to the south of the Balearics, running in close to Majorca. If for some reason we had to put into port I wanted a port to be handy, and Palma was the nearest.

All that night and all the next morning we ran under power. There was no wind nor was there any sign that there was ever going to be any wind ever again. The sky was an immaculate blue echoing the waveless sea and I felt like hell. With no wind a sailing boat is helpless, and what would we do when the fuel ran out?

I discussed it with Coertze. "I'm inclined to put in to Palma," I said. "We can fill up there."

He threw a cigarette stub over the side. "It's a damn' waste of time. We'd be going off course, and what if they keep us waiting round there?"

I said, "It'll be a bigger waste of time if we're left without power. This calm could go on for days."

"I've been looking at the Mediterranean Pilot," he said. "It says the percentage of calms at this time of year isn't high."

"You can't depend on that—those figures are just averages. This could go on for a week."

He sighed. "You're the skipper," he said. "Do the best you can."

So I altered course to the north and we ran for Palma. I checked on the fuel remaining and doubted if we'd make it—but we did. We motored into the yacht harbour at Palma with the engine coughing on the last of the fuel. As we approached the mooring jetty the engine expired and we drifted the rest of the way by momentum.

It was then I looked up and saw Metcalfe.

11

We cut the Customs formalities short by saying that we weren't going ashore and that we had only come in for fuel. The Customs officer commiserated with us on the bad sailing weather and said he would telephone for a chandler to come down and see to our needs.

That left us free to discuss Metcalfe. He hadn't said anything—he had just regarded us with a gentle smile on his lips and then had turned on his heel and walked away.

Coertze said, "He's cooking something up."

"Nothing could be more certain," I said bitterly. "Will we never get these bastards off our backs?"

"Not while we've got four tons of gold under our feet," said Coertze. "It's like a bloody magnet."

I looked forward at Walker sitting alone on the foredeck. There was the fool who, by his loose tongue and his stupidities, had brought the vultures down on us. Or perhaps not—men like Metcalfe and Torloni have keen noses for gold. But Walker hadn't helped.

Francesca said, "What do you think he will do?"

"My guess is a simple act of piracy," I said. "It'll appeal to his warped sense of humour to do some Spanish Main stuff."

I lay on my back and looked at the sky. The club burgee at the masthead was lifting and fluttering in a light breeze. "And look at that," I said. "We've got a wind, dammit."

"I said we shouldn't have come in here," grumbled Coertze. "We'd have had the wind anyway, and Metcalfe wouldn't have spotted us."

I considered Metcalfe's boat and his radar—especially the radar. "No," I said. "It wouldn't have made any difference. He's probably known just where to put his hand on us ever since we left Italy." I made a quick calculation on the basis of a 15-mile radar range. "He can cover 700 square miles of sea with one pass of his radar. That Fairmile has probably been hovering hull-down on the horizon keeping an eye on us. We'd never spot it."

"Well, what do we do now?" asked Francesca.

"We carry on as usual," I said. "There's not much else we can do. But I'm certainly not going to hand the gold to Mr. Bloody Metcalfe simply because he shows up and throws a scare into us. We carry on and hope for the best."

We refuelled and topped up the water tanks and were on our way again before nightfall. The sun was setting as we passed Cabo Figuera and I left the helm to Francesca and went below to study the chart. I had a plan to fox Metcalfe—it probably wouldn't work but it was worth trying.

As soon as it was properly dark I said to Francesca, "Steer course 180 degrees."

"South?" she said in surprise.

"That's right—south." To Coertze I said, "Do you know what that square gadget half-way up the mast is for?"

"*Nee, man,* I've never worried about it."

"It's a radar reflector," I said. "A wooden boat gives a bad radar reflection so we use a special reflector for safety—it gives a nice big blip on a screen. If Metcalfe has been following us he must have got used to that blip by now—he can probably identify us sight unseen, just from the trace on the screen. So we're going to take the reflector away. He'll still get an echo but it'll be different, much fainter."

I fastened a small spanner on a loop round my wrist and clipped a lifeline on to my safety belt and began to climb the mast. The reflector was bolted on to the lower spreaders and it was an uneasy job getting it down. *Sanford* was doing her new style dot-and-carry-one, and following the old-time sailor maxim of "one hand for yourself and one hand for the ship" it was not easy to unfasten those two bolts. The trouble was that the bolts started to turn as well as the nuts, so I was getting nowhere fast. I was up the mast for over forty-five minutes before the reflector came free.

I got down to the deck, collapsed the reflector for stowage and said to Coertze, "Where's Walker?"

"Dossing down; it's his watch at midnight."

"I'd forgotten. Now we change the lights." I went below to the chart table. I had a white light at the masthead visible all round which was coupled to a Morse key for signalling. I tied the key down so that the light stayed on all the time.

Then I called up to Coertze, "Get a lantern out of the fo'c'sle and hoist it in the rigging."

He came below. "What's all this for?"

I said, "Look, we're on the wrong course for Tangier—it's wasting time but it can't be helped because anything that puts Metcalfe off his stroke is good for us. We've altered our radar trace but Metcalfe might get suspicious and come in for a look at us, anyway. So we're festooned with lights in the usual sloppy Spanish fisherman fashion. We're line fishing and he won't see otherwise—not at night. So he just may give us the go-by and push off somewhere else."

"You're a tricky bastard," said Coertze admiringly.

"It'll only work once," I said. "At dawn we'll change course for Tangier."

III

The wind got up during the night and we handed the light weather sails so that *Sanford* developed a fair turn of speed. Not that it helped much; we weren't making an inch of ground in the direction of Tangier.

At dawn it was blowing force five and we changed course so that the wind was on the quarter and *Sanford* began to stride out, her lee rail under and the bow wave showing white foam. I checked the log and saw that she was doing seven knots, which was close to her limit under sail. We were doing all right at last—on the right course for Tangier and travelling fast.

We kept a close watch on the horizon for Metcalfe but saw nothing. If he knew where we were he wasn't showing his hand. I didn't know whether to be glad or sorry about that; I would be glad if my stratagem had deceived him, but if it hadn't then I wanted to know about it.

The fresh breeze held all day and even tended to increase towards nightfall. The waves became larger and foam-crested, breaking every now and then on *Sanford's* quarter. Every time that happened she would shudder and shake herself free to leap forward again. I estimated that the wind was now verging on force six and, as a prudent seaman, I should have been thinking of taking a reef in the mainsail, but I wanted to press on—there was not much time left, and less if we had to tangle with Metcalfe.

I turned in early, leaving Walker at the helm, and before I went to sleep I contemplated what I would do if I were Metcalfe. We had to go through the Straits of Gibraltar—the whole Mediterranean was a funnel with the Straits forming the spout. If Metcalfe took station there his radar could cover the whole channel from shore to shore.

On the other hand, the Straits were busy waters, so he'd have to zig-zag to check dubious boats visually. Then again, if he was contemplating piracy, it would be dangerous to try it where it could be spotted easily—there were some very fast naval patrol boats at Gibraltar and I didn't think that even Metcalfe would have the nerve to tackle us in daylight.

So that settled that—we would have to run through the Straits in daylight.

If—and I was getting tired of all these ifs—if he didn't nobble us before or after the Straits. I hazily remembered a case of piracy just outside Tangier in 1956—two groups of smugglers had tangled with each other and one of the boats had been burned. Perhaps he wouldn't want to leave it as long as that; we would be close to home and we might give him the slip after all—once we were in the yacht harbour there wouldn't be a damn' thing he could do. No, I didn't think he would leave it as late as that.

But before the Straits? That was a different kettle of fish and that depended on another "if." If we had given him the slip on leaving Majorca—if he didn't know where we were now—then we might have a chance. But if he did know where we were, then he could close in anytime and put a prize crew aboard. If—yet another if—the weather would let him.

As I drifted off to sleep I blessed the steadily rising wind which added wings to *Sanford* and which would make it impossible for the Fairmile to come alongside.

IV

Coertze woke me up. "The wind's getting stronger; I think we should change sail or something." He had to shout above the roar of the wind and the sea.

I looked at my watch as I pulled on my oilies; it was two o'clock and I had had six hours' sleep. *Sanford* was bucking a bit and I had a lot of trouble putting my trousers on. A sudden lurch sent me across the cabin and I carommed into the berth in which Francesca slept.

"What is wrong?" she asked.

"Nothing," I said. "Everything is fine; go back to sleep."

"You think I can sleep in this?"

I grinned. "You'll soon get used to it. It's blowing up a bit, but nothing to worry about."

I finished dressing and went up into the cockpit. Coertze was right; we should do something about taking in sail. The wind was blowing at a firm force seven—what old-time sailormen referred to contemptuously as a "yachtsman's gale" and what Admiral Beaufort temperately called a "strong wind."

Tattered clouds fled across the sky, making a baffling

alteration of light and shade as they crossed the moon. The seas were coming up in lumps and the crests were being blown away in streaks of foam. *Sanford* was plunging her head into the seas and every time this happened she would stop with a jerk, losing speed. A reduction of sail would hold her head up and help her motion, so I said to Coertze, my voice raised in a shout, " You're right; I'll reef her down a bit. Hold her as she is."

I snapped a lifeline on to my safety belt and went forward along the crazily shifting deck. It took half an hour to take in two rolls round the boom of the mainsail and to take in the jib, leaving the foresail to balance her head. As soon as I had handed the jib I could feel the difference in motion; *Sanford* rode more easily and didn't ram her bows down as often.

I went back to the cockpit and asked Coertze, " How's that?"

" Better," he shouted. " She seems to be going faster, though."

" She is; she's not getting stopped."

He looked at the piled-up seas. " Will it get worse than this?"

" Oh, this is not so bad," I replied. " We're going as fast as we can, which is what we want." I smiled, because from a small boat everything looked larger than life and twice as dangerous. However I hoped the weather wouldn't worsen; that would slow us down.

I stayed with Coertze for a time to reassure him. It was nearly time for my watch, anyway, and there was no point in going back to sleep. After some time I slipped down into the galley and made some coffee—the stove was rocking crazily in its gimbals and I had to clamp the coffee-pot, but I didn't spill a drop.

Francesca was watching me from her berth and when the coffee was ready I beckoned to her. If she came to the coffee instead of vice versa there was less chance of it spilling. We wedged ourselves in between the galley bench and the companionway, sipping the hot coffee and talking about the weather.

She smiled at me. " You like this weather, don't you?"

" It's fine."

" I think it's a little frightening."

" There's nothing to be frightened of," I said. " Or rather, only one thing."

" What is that?"

" The crew," I said. " You see, small boat design has reached the point of perfection just about, as far as seaworthiness is concerned. A boat like this can take any weather safely if she's handled right—and I'm not saying this because I designed and built her—it applies to any boat of this general type. It's the crew that fails, rather than the boat. You get tired and then you make a mistake—and you only have to make one mistake—you can't play about with the sea."

" How long does it take before the crew gets as tired as that?"

" We're all right," I said cheerfully. " There are enough of us, so that we can all get our sleep, so we can last indefinitely. It's the single-handed heroes who have the trouble."

" You're very reassuring," she said, and got up to take another cup from the shelf. " I'll take Coertze some coffee."

" Don't bother; it'll only get full of salt spray, and there's nothing worse than salted coffee. He'll be coming below in a few minutes—it's my watch."

I buttoned my oilies and tightened the scarf round my neck. " I think I'll relieve him now; he shouldn't really be up there in this weather with that hole in his shoulder. How's it doing, anyway?"

" Healing nicely," she said.

" If he had to have a hole in him he couldn't have done better than that one," I said. " Six inches lower and he'd have been plugged through the heart."

She said, " You know, I'm changing my mind about him. He's not such a bad man."

" A heart of gold beneath that rugged exterior?" I queried, and she nodded. I said, " His heart is set on gold, anyway. We may have some trouble with him if we avoid Metcalfe—don't forget his history. But give the nice man some coffee when he comes below."

I went up into the cockpit and relieved Coertze. " There's coffee for you," I shouted.

" Thanks, just what I need," he answered and went below.

Sanford continued to eat up the miles and the wind continued to increase in force. Good and bad together. I still held on to the sail I had, but when Walker came to relieve me in a cloudy and watery dawn I took in another roll of the mainsail before I went below for breakfast.

Just before I descended the companionway, Walker said, " It'll get worse."

I looked at the sky. " I don't think so; it rarely gets worse than this in the Mediterranean."

He shrugged. "I don't know about the Mediterranean, but I have a feeling it'll get worse, that's all."

I went below feeling glum. Walker had previously shown an uncanny ability to detect changes in the weather on no visible evidence. He had displayed this weather sense before and had invariably been proved right. I hoped he was wrong this time.

He wasn't!

I couldn't take a noonday sight because of thick cloud and bad visibility. Even if I could have seen the sun I doubt if I could have held a sextant steady on that reeling deck. The log reading was 152 miles from noon to noon, *Sanford's* best run ever.

Shortly after noon the wind speed increased greatly to force eight verging on force nine—a strong gale. We handed the mainsail altogether and set the trysail, a triangular handkerchief-sized piece of strong canvas intended for heavy weather. The foresail we also doused with difficulty—it was becoming very dangerous to work on the foredeck.

The height of the waves had increased tremendously and they would no sooner break in a white crest than the wind would tear the foam away to blow it in ragged streaks across the sea. Large patches of foam were beginning to form until the sea began to look like a giant washtub into which someone had emptied a few thousand tons of detergent.

I gave orders that no one should go on deck but the man on watch and that he should wear a safety line at all times. For myself, I got into my berth, put up the bunkboards so that I wouldn't be thrown out, and tried unsuccessfully to read a magazine. But I kept wondering if Metcalfe was out in this sea. If he was, I didn't envy him, because a power boat does not take heavy weather as kindly as a sailing yacht, and he must be going through hell.

Things got worse later in the afternoon so I decided to heave-to. We handed the trysail and lay under bare poles abeam to the seas. Then we battened down the hatches and all four of us gathered in the main cabin chatting desultorily when the noise would allow us.

It was about this time that I started to worry about sea room. As I had been unable to take a sight I didn't know our exact position—and while dead reckoning and log readings were all very well in their way, I was beginning to become perturbed. For we were now in the throat of the funnel between Almeria in Spain and Morocco. I knew we were safe enough from being wrecked on the mainland, but just about here was a fly-speck of an island called Alboran which could be the ruin of us if we ran into it in this weather.

I studied the Mediterranean Pilot. I had been right when I said that this sort of weather was not common in the Mediterranean, but that was cold comfort. Evidently the Clerk of the Weather hadn't read the Mediterranean Pilot—the old boy was certainly piling it on.

At five o'clock I went on deck for a last look round before nightfall. Coertze helped me take away the batten boards from the companion entrance and I climbed into the cockpit. It was knee-deep in swirling water despite the three two-inch drains I had built into it; and as I stood there, gripping a stanchion, another boiling wave swept across the deck and filled the cockpit.

I made a mental note to fit more cockpit drains, then looked at the sea. The sight was tremendous; this was a whole gale and the waves were high, with threatening overhanging crests. As I stood there one of the crests broke over the deck and *Sanford* shuddered violently. The poor old girl was taking a hell of a beating and I thought I had better do something about it. It would mean at least one man in the cockpit getting soaked and miserable and frightened and I knew that man must be me—I wouldn't trust anyone else with what I was about to do.

I went back below. "We'll have to run before the wind," I said. "Walker, fetch that coil of 4-inch nylon rope from the fo'c'sle. Kobus, get into your oilskins and come with me."

Coertze and I went back into the cockpit and I unlashed the tiller. I shouted, "When we run her downwind we'll have to slow her down. We'll run a bight of rope astern and the drag will help."

Walker came up into the cockpit with the rope and he made one end fast to the port stern bitts. I brought *Sanford* downwind and Coertze began to pay the rope over the stern. Nylon, like hemp and unlike manila, floats, and the loop of rope acted like a brake on *Sanford's* wild rush.

Too much speed is the danger when you're running before a gale; if you go too fast then the boat is apt to trip just like a man who trips over his own feet when running. When that happens the boat is likely to capsize fore-and-aft—the bows dig into the sea, the stern comes up and the boat somersaults. It happened to *Tzu Hang* in the Pacific and it happened to Erling Tambs' *Sandefjord* in the Atlantic when he lost a man. I didn't want it to happen to me.

Steering a boat in those conditions was a bit hair-raising.

The stern had to be kept dead in line with the overtaking wave and, if you got it right, then the stern rose smoothly and the wave passed underneath. If you were a fraction out then there was a thud and the wave would break astern; you would be drenched with water, the tiller would nearly be wrenched from your grasp and you would wonder how much more of that treatment the rudder would take.

Coertze had paid out all the nylon, a full forty fathoms, and *Sanford* began to behave a little better. The rope seemed to smooth the waters astern and the waves did not break as easily. I thought we were still going a little too fast so I told Walker to bring up some more rope. With another two lengths of twenty-fathom three-inch nylon also streamed astern I reckoned we had cut *Sanford's* speed down to three knots.

There was one thing more I could do. I beckoned to Coertze and put my mouth close to his ear. "Go below and get the spare can of diesel oil from the fo'c'sle. Give it to Francesca and tell her to put half a pint at a time into the lavatory, then flush it. About once every two or three minutes will do."

He nodded and went below. The four-gallon jerrycan we kept as a spare would now come in really useful. I had often heard of pouring oil on troubled waters—now we would see if there was anything in it.

Walker was busy wrapping sailcloth around the ropes streamed astern where they rubbed on the taffrail. It wouldn't take much of this violent movement to chafe them right through, and if a rope parted at the same moment that I had to cope with one of those particularly nasty waves which came along from time to time then it might be the end of us.

I looked at my watch. It was half past six and it looked as though I would have a nasty and frightening night ahead of me. But I was already getting the hang of keeping *Sanford* stern on to the seas and it seemed as though all I would need would be concentration and a hell of a lot of stamina.

Coertze came back and shouted, "The oil's going in."

I looked over the side. It didn't seem to be making much difference, though it was hard to tell. But anything that *could* make a difference I was willing to try, so I let Francesca carry on.

The waves were big. I estimated they were averaging nearly forty feet from trough to crest and *Sanford* was behaving like a roller-coaster car. When we were in a trough the waves looked frighteningly high, towering above us with threatening crests. Then her bows would sink as a wave took her astern until it seemed as if she was vertical and going to dive straight to the bottom of the sea. The wave would lift her to the crest and then we could see the storm-tattered sea around us, with spume being driven from the waves horizontally until it was difficult to distinguish between sea and air. And back we would go into a trough with *Sanford's* bows pointing to the skies and the monstrous waves again threatening.

Sometimes, about four times in an hour, there was a freak wave which must have been caused by one wave catching up with another, thus doubling it. These freaks I estimated at sixty feet high—higher than *Sanford's* mast!—and I would have to concentrate like hell so that we wouldn't be pooped.

Once—just once—we were pooped, and it was then that Walker went overboard. We were engulfed in water as a vast wave broke over the stern and I heard his despairing shout and saw his white face and staring eyes as he was washed out of the cockpit and over the side.

Coertze reaction was fast. He lunged for Walker—but missed. I shouted, " Safety line—pull him in."

He brushed water out of his eyes and yelled, " Wasn't wearing one."

" The damned fool," I thought. I think it was a thought—I might even have yelled it. Coertze gave a great shout and pointed aft and I turned and could see a dark shape rolling in the boiling waters astern and I saw white hands clutching the nylon rope. They say a drowning man will clutch at a straw—Walker was lucky—he had grabbed at something more substantial, one of the drag ropes.

Coertze was hauling the rope in fast. It couldn't have been easy with the drag of Walker in the water pulling on his injured shoulder, but he was hauling just as fast as though the rope was free. He pulled Walker right under the stern and then belayed the rope.

He shouted to me, " I'm going over the counter—you'll have to sit on my legs."

I nodded and he started to crawl over the counter stern to where Walker was still tightly gripping the rope. He slithered aft and I got up from my seat and hoisted myself out of the cockpit until I could sit on his legs. In the violent motion of the storm it was only my weight that kept Coertze from being hurled bodily into the sea.

Coertze grasped the rope and heaved, his shoulders writhing with the effort. He was lifting the dead weight of Walker five feet—the distance from the taffrail to the surface of the water. I hoped to God that Walker could hold on. If he let go then, not only would he be lost himself but the sudden release of tension would throw Coertze off balance and he would not have a hope of saving himself.

Walker's hands appeared above the taffrail and Coertze took a grip on the cuff of his coat. Then I looked aft and yelled, "Hang on, for God's sake!"

One of those damnable freak seas was bearing on us, a terrifying monster coming up astern with the speed of an express train. *Sanford's* bows sank sickeningly and Coertze gave Walker another heave, and grasped him by the scruff of the neck, pulling him on to the counter.

Then the wave was upon us and away as fast as it had come. Walker tumbled into the bottom of the cockpit, unconscious or dead, I couldn't tell which, and Coertze fell on top of him, his chest heaving with the strain of his exertions. He lay there for a few minutes, then bent down to loosen Walker's iron grip on the rope.

As he prised the fingers away, I said, "Take him below—and you'd better stay there yourself for a while."

A great light had just dawned on me but I had no time to think about it just then—I had to get that bight of rope back over the stern while still keeping a grasp of the tiller and watching the next sea coming up.

It was nearly an hour before Coertze came back—a lonely and frightening hour during which I was too busy to think coherently about what I had seen. The storm seemed to be building up even more strongly and I began to have second thoughts about what I had told Francesca about the sea-worthiness of small boats.

When he climbed into the cockpit he took over Walker's job of looking after the stern ropes, giving me a grin as he settled down. "Walker's O.K.," he bawled. "Francesca's looking after him. I pumped the water out of him—the bilges must be nearly full." He laughed and the volume of his great laughter seemed to overpower the noise of the gale.

I looked at him in wonder.

V

A Mediterranean gale can't last; there is not the power of a huge ocean to draw upon and a great wind soon dies. At four the next morning the storm had abated enough for me to hand over the tiller to Coertze and go below. When I sat on the settee my hands were shaking with the sudden release of tension and I felt inexpressibly weary.

Francesca said, "You must be hungry; I'll get you something to eat."

I shook my head. "No, I'm too tired to eat—I'm going to sleep." She helped me take off my oilies, and I said, "How's Walker?"

"He's all right; he's asleep in the quarter berth."

I nodded slowly—Coertze had put Walker into his own berth. That fitted in, too.

I said, "Wake me in two hours—don't let me sleep any longer. I don't want to leave Coertze alone too long," and I fell on to my berth and was instantly asleep. The last thing I remembered was a fleeting vision of Coertze hauling Walker over the stern by the scruff of the neck.

Francesca woke me at six-thirty with a cup of coffee which I drank gratefully. "Do you want something to eat?" she asked.

I listened to the wind and analysed the motion of *Sanford*. "Make breakfast for all of us," I said. "We'll heave to and have a rest for a bit. I think the time has come for a talk with Coertze, anyway."

I went back into the cockpit and surveyed the situation. The wind was still strong but not nearly as strong as it had been, and Coertze had hauled in the two twenty-fathom ropes and had coiled them neatly. I said, "We'll heave to now; it's time you had some sleep."

He nodded briefly and we began to haul in the bight of rope. Then we lashed the tiller and watched *Sanford* take position broadside on to the seas—it was safe now that the wind had dropped. When we went below Francesca was in the galley making breakfast. She had put a damp cloth on the cabin table to stop things sliding about and Coertze and I sat down.

He started to butter a piece of bread while I wondered how to go about what I was going to say. It was a difficult question I was going to broach and Kobus had such a thorny character that I didn't know how he would take it. I said, "You know, I never really thanked you properly for pulling me out of the mine—you know, when the roof caved in."

He munched on the bread and said, with his mouth full, "*Nee, man,* it was my own fault, I told you that before. I should have shored the last bit properly."

"Walker owes you his thanks, too. You saved his life last night."

He snorted. "Who cares what he thinks."

I said carefully, getting ready to duck, "Why did you do it, anyway? It would have been worth at least a quarter of a million *not* to pull him out."

Coertze stared at me, affronted. His face reddened with anger. "Man, do you think I'm a bloody murderer?"

I had thought so at one time but didn't say so. "And you didn't kill Parker or Alberto Corso or Donato Rinaldi?"

His face purpled. "Who said I did?"

I cocked my thumb at the quarter berth where Walker was still asleep. "He did."

I thought he would burst. His jaws worked and he was literally speechless, unable to say a damn' thing. I said, "According to friend Walker, you led Alberto into a trap on a cliff and then pushed him off; you beat in the head of Donato; you shot Parker in the back of the head when you were both in action against the Germans."

"The lying little bastard," ground out Coertze. He started to get up. "I'll ram those lies down his bloody throat."

I held up my hand. "Hold on—don't go off half-cocked. Let's sort it out first; I'd like to get your story of what happened at that time. You see, what happened last night has led me to reconsider a lot of things. I wondered why you should have saved Walker if you're the man he says you are. I'd like to get at the truth for once."

He sat down slowly and looked down at the table. At last he said, "Alberto's death was an accident; I tried to save him, but I couldn't."

"I believe you—after last night."

"Donato I know nothing about. I remember thinking that there was something queer about it, though. I mean, why should Donato go climbing for fun? He had enough of that the way the Count sent us all over the hills."

"And Parker?"

"I couldn't have killed Parker even if I'd wanted to," he said flatly.

"Why not?"

Slowly he said, "We were with Umberto doing one of the usual ambushes. Umberto split the force in two—one group on one side of the valley, the other group on the other side. Parker and Walker were with the other group. The ambush was a flop, anyway, and the two parties went back to camp separately. It was only when I got back to camp that I heard that Parker had been killed."

He rubbed his chin. "Did you say that Walker told you that Parker had been shot in the back of the head?"

"Yes."

He looked at his hands spread out on the table. "Walker could have done it, you know. It would be just like him."

"I know," I said. "You told me once that Walker had got you into trouble a couple of times during the war. When exactly did that happen? Before you buried the gold or afterwards?"

He frowned in thought, casting his mind back to faraway days. He said, "I remember once when Walker pulled some men away from a ditch when he shouldn't have. He was acting as a mesenger for Umberto and said he'd misunderstood the instruction. I was leading a few chaps at the time and this left my flank wide open." His eyes darkened. "A couple of the boys copped it because of that and I nearly got a bayonet in my rump." His face twisted in thought. "It was *after* we buried the gold."

"Are you sure?"

"I'm certain. We only joined Umberto's crowd after we'd buried the gold."

I said softly, "Maybe he could shoot Parker in the back of the head. Maybe he could beat in the back of Donato's head with a rock and fake a climbing accident. But maybe he was too scared of *you* to come at you front or rear—you're a bit of an awesome bastard at times, you know. Maybe he tried to arrange that the Germans should knock you off."

Coertze's hands clenched on the table. I said, "He's always been afraid of you—he still is."

"*Magtig,* but he has reason to be," he burst out. "Donato got us out of the camp. Donato stayed with him on the hillside while the Germans were searching." He looked at me with pain in his eyes. "What kind of a man is it who can do such a thing?"

"A man like Walker," I said. "I think we ought to talk to him. I'm getting eager to know what he's arranged for me and Francesca."

Coertze's lips tightened. "*Ja,* I think we wake him up now out of that *lekker slaap.*"

He stood up just as Francesca came in loaded with bowls. She saw Coertze's face and paused uncertainly. "What's the matter?"

I took the bowls from her and put them in the fiddles. "We're just going to have a talk with Walker," I said. "You'd better come along."

But Walker was already awake and I could see from his expression that he knew what was coming. He swung himself from the berth and tried to get away from Coertze, who lunged at him.

"Hold on," I said, and grabbed Coertze's arm. "I said we're going to talk to him."

The muscles bunched in Coertze's arm and then relaxed and I let go. I said to Walker, "Coertze thinks you're a liar—what do you say?"

His eyes shifted and he gave Coertze a scared glance, then he looked away. "I didn't say he killed anybody. I didn't say that."

"No, you didn't," I agreed. "But you damn' well implied it."

Coertze growled under his breath but said nothing, apparently content to let me handle it for the moment. I said, "What about Parker? You said that Coertze was near him when he was shot—Coertze said he wasn't. What about it?"

"I didn't say that either," he said sulkily.

"You *are* a damned liar," I said forcibly. "You said it to me. I've got a good memory even if you haven't. I warned you in Tangier what would happen if you ever lied to me, so you'd better watch it. Now I want the truth—was Coertze near Parker when he was killed?"

He was silent for a long time. "Well, was he?" I demanded.

He broke. "No, he wasn't," he cried shrilly. "I made that up. He wasn't there; he was on the other side of the valley."

"Then who killed Parker?"

"It was the Germans," he cried frantically. "It was the Germans—I told you it was the Germans."

I suppose it was too much to expect him to confess to murder. He would never say outright that he had killed Parker and Donato Rinaldi—but his face gave him away. I had no intention of sparing him anything, so I said to Coertze, "He was responsible for Torloni's attack."

Coertze grunted in surprise. "How?"

I told him about the cigarette case, then said to Walker, "Coertze saved your life last night, but I wish to God he'd let you drown. Now I'm going to leave him down here alone with you and he can do what he likes."

Walker caught my arm. "Don't leave me," he implored. "Don't let him get at me." What he had always feared was now about to happen—there was now no one between him and Coertze. He had blackened Coertze in my eyes so that he would have an ally to fight his battles, but now I was on

Coertze's side. He feared physical violence—his killings had been done from ambush—and Coertze was the apotheosis of violence.

"Please," he whimpered, "don't leave." He looked at Francesca with a passionate plea in his eyes. She turned aside without speaking and went up the companionway into the cockpit. I shook off his hand and followed her, closing the cabin hatch.

"Coertze will kill him," she said in a low voice.

"Hasn't he the right?" I demanded. "I don't believe in private executions as a rule, but this is one time I'm willing to make an exception."

"I'm not thinking of Walker," she said. "It will be bad for Coertze. No one can kill a man like that and be the same after. It will be bad for his . . . his spirit."

I said, "Coertze will do what he has to do."

We lapsed into silence, just looking at the lumpy sea, and I began to think of the boat and what we had to do next.

The cabin hatch opened and Coertze came into the cockpit. There was a baffled expression on his face and he said in a hoarse voice, "I was going to kill the little bastard. I was going to hit him—I did hit him once. But you can't hit anyone who won't fight back. You can't, can you?"

I grinned and Francesca laughed joyously. Coertze looked at us and his face broke into a slow smile. "But what are we going to do with him?" he asked.

"We'll drop him at Tangier and let him shift for himself," I said. "We'll give him the biggest scare any man's ever had."

We were sitting grinning at each other like a couple of happy fools when Francesca said sharply, "Look!"

I followed the line of her outstretched arm. "Oh, no!" I groaned. Coertze looked and cursed.

Coming towards us through the tossing seas and wallowing atrociously was the Fairmile.

IX. SANFORD

I looked at it bitterly. I had been certain that Metcalfe must have lost us in the storm—he had the luck of the devil. He hadn't found us by radar either, because the storm had made a clean sweep of the Fairmile's upperworks—his radar antenna was gone, as also was the radio mast and the short derrick. It could only have been by sheer luck that he had stumbled upon us.

I said to Coertze, "Get below and start the engine. Francesca, you go below, too, and stay there."

I looked across at the Fairmile. It was about a mile away and closing at about eight knots—a little over five minutes to make what futile preparations we could. I had no illusions about Metcalfe. Torloni had been bad enough but all he knew was force—Metcalfe used his brains.

The seas were bad. When the wind had blown it exerted a discipline on the sea and the waves were regular and uniform, With the wind gone the sea was chaotic; ugly pyramidical lumps of water would heave up and disappear and *Sanford* pitched and rolled alarmingly.

The Fairmile was in no better shape, either. She staggered and wallowed as unexpected waves hit her and I could imagine the tumult inside that hull. She was an old boat, being war surplus, and her hull must have deteriorated over the years despite the care Metcalfe had lavished on her. Then there was the fact that when she was built her life expectancy was about five years, and war-time materials weren't noted for their excessive quality.

I had the sudden idea that she couldn't move any faster, and that Metcalfe was driving her as fast as he dared in those heavy seas. Her engines were fine for twenty-six knots in calm water but if she was driven at much more than eight knots now she would be in danger of falling apart. Metcalfe might risk a lot for the gold, but he wouldn't risk that.

As I heard the engine start I opened the throttle wide and turned *Sanford* away from the Fairmile. We had a biggish engine and I could still get seven knots out of *Sanford,* even punching against these seas. Our five minutes' grace was now stretched to an hour, and maybe in that hour I'd get another bright idea.

Coertze came up and I handed the tiller over to him, and went below. I didn't bother to tell him what to do—it was obvious. I opened the locker under my berth and took out the Schmeisser machine pistol and all the magazines. Francesca looked at me from the settee. "Must you do that?" she asked.

"I'll not shoot unless I have to," I said. "Not unless they start shooting first." I looked around. "Where's Walker?"

"He locked himself in the fo'c'sle. He's frightened of Coertze."

"Good. I don't want him underfoot now," I said, and went back to the cockpit.

Coertze looked incredulously at the machine pistol. "Where the hell did you get that?"

"From the tunnel," I said. "I hope it works—this ammo is damned old."

I put one of the long magazines into the butt and clipped the shoulder rest into place. I said, "You'd better get your Luger; I'll take the helm."

He smiled sourly. "What's the use? You threw all the bullets away."

"Damn! Wait a minute, though; there's Torloni's gun. It's in the chart table drawer."

He went below and I looked back at the Fairmile. As I thought, Metcalfe didn't increase his speed when we turned away. Not that it mattered—he had the legs of us by about a knot and I could see that he was perceptibly closer.

Coertze came back with the pistol stuck in his trouser waistband. He said, "How long before he catches up?"

"Less than an hour," I said. I touched the Schmeisser. "We don't shoot unless he does—and we don't shoot to kill."

"Will he shoot to kill?"

"I don't know," I said. "He might."

Coertze grunted and pulled out the gun and began to examine the action.

We fell into silence; there was nothing much to talk about, anyway. I ruminated on the firing of a sub-machine-gun. It had been a long time since I had fired one and I began to go over the training points that had been drilled into me by a red-faced sergeant. The big thing was that the recoil lifted the muzzle and if you didn't consciously hold it down most of your fire would be wasted in the air. I tried to think of other things I had learned but I couldn't think of anything else so that fragment of information would have to do.

After a while I said to Coertze, "I could do with some coffee."

"That's not a bad idea," he said, and went below. An Afrikaner will never refuse the offer of coffee; their livers are tanned with it. In five minutes he was back with two steaming mugs, and said, "Francesca wants to come up."

I looked back at the Fairmile. "No," I said briefly.

We drank the coffee, spilling half of it as *Sanford* shuddered to a particularly heavy sea, and when we had finished the Fairmile was within a quarter of a mile and I could see Metcalfe quite clearly standing outside the wheelhouse.

I said, "I wonder how he's going to go about it. He can't

board us in this sea, there's too much danger of ramming us. How would you go about it, Kobus?"

"I'd lay off and knock us off with a rifle," he grunted. "Just like at a shooting gallery. Then when the sea goes down he can board us without a fight."

That seemed reasonable but it wouldn't be as easy as in a shooting gallery—metal ducks don't shoot back. I handed the tiller to Coertze. "We may have to do a bit of fancy manœuvring," I said. "But you'll handle her well enough without sail. When I tell you to do something, you do it damn' quick." I picked up the Schmeiser and held it on my knee. "How many rounds are there in that pistol?"

"Not enough," he said. "Five."

At last the Fairmile was only a hundred yards away on the starboard quarter and Metcalfe came out of the wheelhouse carrying a Tannoy loud-hailer. His voice boomed across the water. "What are you running away for? Don't you want a tow?"

I cupped my hands around my mouth. "Are you claiming salvage?" I asked sardonically.

He laughed. "Did the storm do any damage?"

"None at all," I shouted. "We can get to port ourselves." If he wanted to play the innocent I was prepared to go along with him. I had nothing to lose.

The Fairmile was throttled back to keep pace with us. Metcalfe fiddled with the amplification of the loud-hailer and it whistled eerily. "Hal," he shouted, "I want your boat—and your cargo."

There it was—out in the open as bluntly as that.

The loud-hailer boomed, "If you act peaceable about it I'll accept half, if you don't I'll take the lot, anyway."

"Torloni made the same offer and look what happened to him."

"He was at a disadvantage," called Metcalfe. "He couldn't use guns—I can."

Krupke moved into sight—carrying a rifle. He climbed on top of the deck saloon and lay down just behind the wheelhouse. I said to Coertze, "It looks as though you called that one."

It was bad, but not as bad as all that. Krupke had been in the army; he was accustomed to firing from a steady position even though his target might move. I didn't think he could fire at all accurately from a bouncing platform like the Fairmile.

I saw the Fairmile edging in closer and said to Coertze, "Keep the distance."

Metcalfe shouted, "What about it?"

"Go to hell!"

He nodded to Krupke, who fired immediately. I didn't see where the bullet went—I don't think it hit us at all. He fired again and this time he hit something forward. It must have been metal because I heard a "spaaang" as the bullet ricochetted away.

Coertze dug me in the ribs. "Don't look back so that Metcalfe notices you, but I think we're in for some more heavy weather."

I changed position on the seat so that I could look astern from the corner of my eye. The horizon was black with a vicious squall—and it was coming our way. I hoped to God it would hurry.

I said, "We'll have to play for time now."

Krupke fired again and there was a slam astern. I looked over the side and saw a hole punched into the side of the counter. His aim was getting better.

I shouted, "The Krupke not to hole us below the waterline. We might sink and you wouldn't like that."

That held him for a while. I saw him talking to Krupke, making gestures with his hand to indicate a higher elevation. I called urgently for Francesca to come on deck. Those nickel-jacketed bullets would go through *Sanford's* thin planking as though it was tissue paper. She came up just as Krupke fired his next shot. It went high and didn't hit anything.

As soon as Metcalfe saw her he held up his hand and Krupke stopped firing. "Hal, be reasonable," he called. "You've got a woman aboard."

I looked at Francesca and she shook her head. I shouted, "You're doing the shooting."

"I don't want to hurt anybody," pleaded Metcalfe.

"Then go away."

He shrugged and said something to Krupke, who fired again. The bullet hit the gooseneck with a clang. I grinned mirthlessly at Metcalfe's curious morality—according to him it would be my fault if anyone was killed.

I looked astern. The squall was appreciably nearer and coming up fast. It was the last dying kick of the storm and wouldn't last long—just long enought to give Metcalfe the slip, I hoped. I didn't think that Metcalfe had seen it yet; he was too busy with us.

Krupke fired again. There was a thud forward and I knew

the bullet must have gone through the main cabin. I had brought Francesca up just in time.

I was beginning to worry about Krupke. In spite of the difficulties of aiming, his shooting was getting better, and even if it didn't, then sooner or later he would get in a lucky shot. I wondered how much ammunition he had.

"Metcalfe," I called.

He held up his hand but not soon enough to prevent Krupke pulling the trigger. The cockpit coaming disintegrated into matchwood just by my elbow. We all ducked low into the cockpit and I looked incredulously at the back of my hand—a two-inch splinter of mahogany was sticking in it.

I pulled it out and shouted, "Hey, hold it! That was a bit too close."

"What do you want?"

I noticed that the Fairmile was crowding us again so I told Coertze to pull out.

"Well?" Metcalfe's voice was impatient.

"I want to make a deal," I shouted.

"You know my terms."

"How do we know we can trust you?"

Metcalfe was uncompromising. "You don't."

I pretended to confer with Coertze. "How's that squall coming up?"

"If you keep stringing him along we might make it."

I turned to the Fairmile. "I'll make a counter-proposition. We'll give you a third—Walker won't be needing his share."

Metcalfe laughed. "Oh, you've found him out at last, have you?"

"What about it?"

"Nothing doing—half or all of it. Make your choice; you're in no position to bargain."

I turned to Coertze. "What do you think, Kobus?"

He rubbed his chin. "I'll go along with anything you say."

"Francesca?"

She sighed. "Do you think this other storm coming up will help?"

"It's not a storm, but it'll help. I think we can lose Metcalfe if we can hold him off for another ten minutes."

"Can we?"

"I think so, but it might be dangerous."

Her lips tightened. "Then fight him."

I looked across at Metcalfe. He was standing by the door of the wheelhouse looking at Krupke who was pointing astern.

He had seen the squall!

I shouted. "We've been having a conference and the general consensus of opinion is that you can still go to hell."

He jerked his hand irritably and Krupke fired again—another miss.

I said to Coertze, "We'll give him another two shots. Immediately after the second, starboard your helm as though you are going to ram him, but for God's sake don't ram him. Get as close as you can and come back on course parallel to him. Understand?"

As he nodded there was another shot from Krupke. That one hit *Sanford* just under the cockpit—Krupke was getting too good.

Metcalfe couldn't know that we had a machine-gun. *Sanford* had been searched many times and machine-guns—even small ones—aren't to be picked up on every street corner in Italy. There was the chance we could give him a fright. I said to Francesca, "When we start to turn get down in the bottom of the cockpit."

Krupke fired again, missed, and Coertze swung the tiller over. It caught Metcalfe by surprise—this was like a rabbit attacking a weasel. We had something like twenty seconds to complete the manœuvre and it worked. By the time he had recovered enough to shout to the helmsman and for the helmsman to respond, we were alongside.

Krupke fired again when he saw us coming but the bullet went wild. I saw him aim at me and looked right into the muzzle of his gun. Then I cut loose with the Schmeisser.

I had only time to fire two bursts. The first one was for Krupke—I must get him before he got me. Two or three rounds broke the saloon windows of the Fairmile and I let the recoil lift the gun. Bullets smashed into the edge of the decking and I saw Krupke reel back with both hands clasped to his face and heard a thin scream.

Then I switched to the wheelhouse and hosed it. Glass flew but I was too late to catch Metcalfe who was already out of sight. The Schmeisser jammed on a defective round and I yelled at Coertze, "Let's get out of here," and he swung the helm over again.

"Where to?" he asked.

"Back to where we came from—into the squall."

I looked back at the Fairmile. Metcalfe was on top of the deck saloon bending over Krupke and the Fairmile was still continuing on her original course. But her bows were swinging from side to side as though there was nobody at the helm. "This might just work," I said.

But after two or three minutes she started to turn and was soon plunging after us. I looked ahead and prayed we could get into the squall in time. I had never before played for dirty weather.

II

It was nip and tuck but we made it. The first gusts hit us when the Fairmile was barely two hundred yards behind, and ten seconds later she was invisible, lost in spearing rain and sea spume.

I throttled back the engine until it was merely ticking over; it would be suicide to try to butt our way through this. It was an angry bit of weather, all right, but it didn't have the sustained ferocity of the earlier storm and I knew it would be over in half an hour or so.

In that short time we had to lose Metcalfe.

I left the tiller to Coertze and stumbled forward to the mast and hoisted the trysail. That would give us leeway and we could pick a course of sorts. I chose to beat to windward; that was the last thing Metcalfe would expect me to do in heavy weather, and I hoped that when the squall had blown out he would be searching to leeward.

Sanford didn't like it. She bucked and pitched more than ever and I cursed the crankiness caused by the golden keel, the cause of all our troubles. I said to Francesca, "You and Kobus had better go below; there's no point in all of us getting soaked to the skin."

I wondered what Metcalfe was doing. If he had any sense he would have the Fairmile lying head to wind with her engines turning just enough to keep position. But he wanted the gold and had guts enough to try anything weird as long as the boat didn't show signs of falling apart under him. He had shown his seamanship by coming through the big storm undamaged —this squall wouldn't hurt him.

Just then *Sanford* lurched violently and I thought for a moment that she was falling apart under *me*. There was a curious feel to the helm which I couldn't analyse—it was like nothing I had felt on a boat before. She lurched again and seemed to sideslip in the water and she swayed alarmingly even when she hadn't been pushed. I leaned on the helm tentatively and she came round with a rush.

Hastily I pulled the other way and she came back fast, overshooting. It was like riding a horse with a loose saddle and I couldn't understand it.

I had a sudden and dreadful thought and looked over the side. It was difficult to make out in the swirl of water but her boot-topping seemed to be much higher out of the water than it should have been, and I knew what had happened.

It was her keel—that goddammed golden keel.

Coertze had warned us about it. He had said that it would be full of flaws and cracks and that it would be structurally weak. *Sanford* had taken a hell of a hammering in the last couple of days and this last squall was the straw that broke the camel's back—or broke the ship's keel.

I looked over the side again, trying to estimate how much higher she was in the water. As near as I could judge three parts of the keel were gone. *Sanford* had lost three tons of ballast and she was in danger of capsizing at any moment.

I hammered on the cabin hatch and yelled at the top of my voice. Coertze popped his head out. "What's wrong?" he shouted.

"Get on deck fast—Francesca, too. The bloody keel's gone. We're going to capsize."

He looked at me blankly. "What the hell do you mean?" His face flushed red as the meaning sank in. "You mean the gold's gone?" he said incredulously.

"For Christ's sake, don't just stand there gaping," I shouted. "Get the hell up here—and get Francesca out of there. I don't know if I can hold her much longer."

He whitened and his head vanished. Francesca came scrambling out of the cabin with Coertze on her heels. *Sanford* was behaving like a crazy thing and I shouted to Coertze, "Get that bloody sail down quick or she'll be over."

He lunged forward along the deck and wasted no time in unfastening the fall of the halyard from the cleat—instead he pulled the knife from his belt and cut it with one clean slice. As soon as the sail came down *Sanford* began to behave a little better, but not much. She slithered about on the surface of the water and it was by luck, not judgment, that I managed to keep her upright, because I had never had that experience before—few people have.

Coertze came back and I yelled, "It's the mast that'll have us over if we're not careful."

He looked up at the mast towering overhead and gave a quick nod. I wondered if he remembered what he had said the first time I questioned him about yachts in Cape Town. He had looked up at the mast of *Estralita* and said, "She'll need to be deep to counter-balance that lot."

The keel, our counter-balance, had gone and the fifty-five-foot mast was the key to *Sanford's* survival.

I pointed to the hatchet clipped to the side of the cockpit. "Cut the shrouds," I shouted.

He seized the hatchet and went forward again and swung at the after starboard shroud. It bounced off the stainless steel wire and I cursed myself for having built *Sanford* so stoutly. He swung again and again and finally the wire parted.

He went on to the forward shroud and I said, "Francesca, I'll have to help him or it may be too late. Can you take the helm?"

"What must I do?"

"I think I've got the hang of it," I said. "She's very tender and you mustn't move the tiller violently. She swings very easily so you must be very gentle in your movements—otherwise it's the same as before."

I couldn't stay with her long before I had to leave the cockpit and release both the backstay runners so that the stays hung loose. The mast now had no support from aft.

I went forward to the bows, clinging on for dear life, and crouched in the bow pulpit, using the marline-spike of my knife on the rigging screw of the forestay. The spike was not designed for the job and kept slipping out of the holes of the body, but I managed to loosen the screws apreciably in spite of being drenched every time *Sanford* dipped her bows. When I looked up I saw a definite curve in the stay to leeward which meant that it was slack.

I looked round and saw Coertze attacking the port shrouds before I bent to loosen the fore topmast stay. When I looked up again the mast was whipping like a fishing rod—but still the damn' thing wouldn't break.

It was only when I tripped over the fore hatch that I remembered Walker. I hammered on the hatch and shouted, "Walker, come out; we're sinking." But I heard nothing from below.

Damning his minuscule soul to hell, I went aft and clattered down the companionway and into the main cabin. I staggered forward, unable to keep my balance in *Sanford's* new and uneasy motion and tried the door to the fo'c'sle. It was locked from the inside. I hammered on it with my fist, and shouted, "Walker, come out; we're going to capsize."

I heard a faint sound and shouted again. Then he called, "I'm not coming out."

"Don't be a damned fool," I yelled. "We're liable to sink at any minute."

"It's a trick to get me out. I know Coertze's waiting for me."

"You bloody idiot," I screamed and hammered on the door again, but it was no use; he refused to answer so I left him there.

As I turned to go, *Sanford* groaned in every timber and I made a mad dash for the companionway, getting into the cockpit just in time to see the mast go. It cracked and split ten feet above the deck and toppled into the raging sea, still tethered by the back and fore stays.

I took the tiller from Francesca and tentatively moved it. *Sanford's* motion was not much better—she still slid about unpredictably—but I felt easier with the top hamper gone. I kicked at a cockpit locker and shouted to Francesca, "Life jackets—get them out."

The solving of one problem led directly to another—the mast in the water was still held fore and aft and it banged rhythmically into *Sanford's* side. Much of that treatment and she would be stove in and we would go down like a stone. Coertze was in the bows and I could see the glint of the hatchet as he raised it for another blow at the forestay. He was very much alive to the danger inherent in the mast.

I struggled into a life jacket while Francesca took the helm, then I grabbed the boathook from the coach roof and leaned over the side to prod the mast away when it swung in again for another battering charge. Coertze came aft and started to cut away the backstays; it was easier to cut them on the deck and within five minutes he had done it, and the mast drifted away and was lost to sight amid the sea spray.

Coertze dropped heavily into the cockpit, his face streaming with salt water, and Francesca gave him a life jacket. We fastened our safety lines and, on a sudden impulse, I battened down the main hatch—if Walker wanted to come out he could still use the fore hatch. I wanted to seal *Sanford*—if she capsized and filled with water she would sink within seconds.

Those last moments of the squall were pretty grim. If we could last them out we might stand a chance. *Sanford* would never sail again, but it might be possible to move her slowly by a judicious use of her engine. For the first time I hoped I had not misled Metcalfe and that he would be standing by.

But the squall had not done with us. A violent gust of wind coincided with a freak sea and *Sanford* tilted alarmingly. Desperately I worked the tiller, but it was too late and she heeled more and more until the deck was at an angle of forty-five degrees.

I yelled, "Hang on, she's going," and in that moment *Sanford* lurched right over and I was thrown into the sea.

I spluttered and swallowed salt water before the buoyancy of the jacket brought me to the surface, lying on my back. Frantically I looked round for Francesca and was relieved when her head bobbed up close by. I grabbed her safety line and pulled until we floated side by side. "Back to the boat," I spluttered.

We hauled on the safety lines and drew ourselves back to *Sanford*. She was lying on her starboard side, heaving sluggishly over the waves, and we painfully crawled up the vertical deck until we could grasp the stanchions of the port safety rail. I looked back over the sea for Coertze, but didn't see him, so I drew myself over the rail and on to the new and oddly shaped upper deck—the port side of *Sanford*.

I helped Francesca over the rail and then I saw Coertze clinging to what was left of the keel—he had evidently jumped the other way. He was clutching a tangle of broken wires—the wires that were supposed to hold the keel together and which had failed in their purpose. I slid down the side and gave him a hand, and soon the three of us were uneasily huddled on the unprotected hull, wondering what the hell to do next.

That last flailing gust of wind had been the squall's final crack of the whip and the wind dropped within minutes to leave the hulk of *Sanford* tossing on an uneasy sea. I looked around hopefully for Metcalfe but the Fairmile wasn't in sight, although she could still come out of the dirty weather left in the wake of the squall.

I was looking contemplatively at the dinghy which was still lashed to the coach roof when Coertze said, "There's still a lot of gold down there, you know." He was staring back at the keel.

"To hell with the gold," I said. "Let's get this dinghy free."

We cut the lashings and let the dinghy fall into the sea—after I had taken the precaution of tying a line to it. It floated upside down, but that didn't worry me—the buoyancy chambers would keep it afloat in any position. I went down the deck and into the sea and managed to right it. Then I took the baler which was still clipped in place and began to bale out.

I had just finished when Francesca shouted, "Metcalfe! Metcalfe's coming."

By the time I got back on top of the hull the Fairmile was quite close, still plugging away at the eight knots which Metcalfe favoured for heavy seas. We weren't trying to get away this time, so it was not long before she was within hailing distance.

Metcalfe was outside the wheelhouse. He bellowed, "Can you take a line?"

Coertze waved and the Fairmile edged in closer and Metcalfe lifted a coil of rope and began to swing it. His first throw was short, but Coertze caught the second and slid down the deck to make the line fast to the stump of the mast. I cut two lengths of line and tied them in loops round the rope Metcalfe had thrown. I said, "We'll go over in the dinghy, pulling ourselves along the line. For God's sake, don't let go of these loops or we might be swept away."

We got into the dinghy and pulled ourselves across to the Fairmile. It wasn't a particularly difficult job but we were cold and wet and tired and it would have been easy to make a mistake. Metcalfe helped Francesca on board and Coertze went next. As I started to climb he threw me a line and said curtly, "Make the dinghy fast; I might need it."

So I made fast and climbed on deck. Metcalfe stepped up to me, his face contorted with rage. He grabbed me by the shoulders with both hands and yelled. "You damn' fool—I told you to make certain of that keel. I told you back in Rapallo."

He began to shake me and I was too tired to resist. My head lolled back and forward like the head of a sawdust doll and when he let me go I just sat down on the deck.

He swung round to Coertze. "How much is left?" he demanded.

"About a quarter."

He looked at the hulk of *Sanford,* a strained expression in his eyes. "I'm not going to lose that," he said. "I'm not going to lose a ton of gold."

He called to the wheelhouse and the Moroccan, Moulay Idriss, came on deck. Metcalfe gave quick instructions in Arabic and then dropped into the dinghy and pulled himself across to *Sanford*. The Arab attached a heavy cable to the line and when Metcalfe got to the hulk he began to pull it across.

Francesca and I were not taking much interest in this. We were exhausted and more preoccupied in being alive and together than with what happened to the gold. Coertze, however, was alive to the situation and was helping the Arab make the cable fast.

Metcalfe came back and said to Coertze, "You were right, there's about a ton left. I don't know how that wreck will behave when it's towed, but we'll try."

As the Fairmile turned and the cable tautened, a watery sun shone out over the heaving sea and I looked back at *Sanford* as she moved sluggishly to the pull. The cockpit was half under water but the fore hatch was still free, and I said, "My God! Walker's still in there!"

Coertze said, "*Magtig,* I'd forgotten him."

He must have been knocked unconscious when *Sanford* capsized—otherwise we would have heard him. Francesca was staring back at *Sanford.* "Look!" she exclaimed. "There—in the cockpit."

The main hatch was being forced open from the inside and I could see Walker's head as he tried to struggle out against the rush of water pouring into the boat. His hands grasped for the cockpit coaming—but it wasn't there—Krupke had shot it away. Then Walker disappeared as the force of the water pushed him back into the cabin.

If he had come out by the fore hatch he would have been safe, but even in death he had to make one of his inevitable mistakes. The main hatch was open, water was pouring into the hull and *Sanford* was sinking.

Metcalfe was in a rage. "The damn' fool," he cried. "I thought you'd got rid of him. He's taking the bloody gold with him."

Sanford was getting low in the water and as she did so, the water poured into her faster. Metcalfe stared at her in despair, his voice filled with fury. "The stupid, bloody idiot," he yelled. "He's bitched things from the start."

It wouldn't be long now—*Sanford* was going fast. The towing cable tightened as she sank lower in the water and the Fairmile went down by the stern as the pull on the cable became greater. *Sanford* gave a lurch as compressed air in the fo'c'sle blew out the forehatch and she began to settle faster as more water poured in through this new opening in her hull.

The downward drag on the stern of the Fairmile was becoming dangerous and Metcalfe took a hatchet from a clip and stood by the cable. He looked back at *Sanford,* his face twitching with indecision, then he brought the hatchet down on the cable with a great swing. It parted with a twang, the loose end snaked away across the sea and the Fairmile bobbed up her stern.

Sanford lurched again and turned over. As she went down and out of sight amid swirling waters a vagrant sunbeam touched her keel and we saw the glint of imperishable gold. Then there was nothing but the sea.

III

Metcalfe's anger was great but, like the squall, soon subsided and he became his usual saturnine self, taking the loss with a philosophical air. "A pity," he said. "But there it is. It's gone and there isn't anything we can do about it now."

We were sitting in the saloon of the Fairmile, on our way to Malaga where Metcalfe was going to drop us. He had given us dry clothing and food and we were all feeling better.

I said, "What will you do now?"

He shrugged. "Tangier is just about played out now the Moroccans are taking over. I think I'll pop down to the Congo—things seem to be blowing up down there."

Metcalfe and a few others like him would be "popping down to the Congo," I thought. Carrion crows flocking together—but he wasn't as bad as some. I said, "I think you've got a few things to explain."

He grinned. "What do you want to know?"

"Well, the thing that's been niggling me is how you got on to us in the first place. What led you to suspect that we were after the gold?"

"Suspect, old boy? I didn't suspect, I *knew*."

"How the devil did you know?"

"It was when I got Walker drunk. He spilled the whole story about the gold, the keel—everything."

"Well, I'm damned." I thought of all the precautions I'd taken to put Metcalfe off the scent; I thought of all the times I'd beaten my brains out to think up new twists of evasion. All wasted—he wasn't fooled at all!

"I thought you'd get rid of him," Metcalfe said. "He was a dead loss all the way through. I thought you'd put him over the side or something like that."

I looked at Coertze, who grinned at me. I said, "He was probably a murderer, too."

"Wouldn't be surprised," agreed Metcalfe airily. "He was a slimy little rat."

That reminded me—I had probably killed a man too. "Where's Krupke?" I asked. "I haven't seen him around."

Metcalfe snickered. "He's groaning in his bunk—he got a faceful of splinters."

I held out the back of my hand. "Well, he did the same to me."

"Yes," said Metcalfe soberly. "But Krupke is probably going to lose an eye."

"Serve him damn' well right," I said viciously. "He won't be too keen to look down rifle sights again."

I hadn't lost sight of the fact that Metcalfe and his crew of ruffians had been doing their damnedest to kill us not many hours before. But there wasn't any advantage in quarrelling with Metcalfe about it—we were on his boat and he was going to put us ashore safely. Irritating him wasn't exactly the best policy just then.

He said, "That machine-gun of yours was some surprise. You nearly plugged me." He pointed to a battered loud-hailer on the sideboard. "You shot that goddamm thing right out of my hand."

Francesca said, "Why were you so solicitous about my husband? Why did you take the trouble?"

"Oh, I felt real bad when I saw Hal slug him," said Metcalfe seriously. "I knew who he was, you see, and I knew he could make a stink. I didn't want anything like that. I wanted Hal to get on with casting the keel and get out of Italy. I couldn't afford to have the police rooting round."

"That's why you tried to hold Torloni, too," I said.

He rubbed his chin. "That was *my* mistake," he admitted. "I thought I could use Torloni without him knowing it. But he's a bad bastard and when he got hold of that cigarette case the whole thing blew up in my face. I just wanted Torloni to keep an eye on you, but that damn' fool, Walker, had to go and give the game away. There was no holding Torloni then."

"So you warned us."

He spread his hands. "What else could I do for a pal?"

"Pal nothing. You wanted the gold out."

He grinned. "Well, what the hell; you got away, didn't you?"

I had bitter thoughts of Metcalfe as the puppet master; he had manipulated all of us and we had danced to his tune. Not quite—one of his puppets had a broken string; if Walker had defeated us, he had also defeated Metcalfe.

I said, "If you hadn't been so obvious about Torloni the keel wouldn't have broken. We had to cast it in a bloody hurry when he started putting the pressure on."

"Yes," said Metcalfe. "And all those damned partisans didn't help, either." He stood up. "Well, I've still got to run this boat." He hesitated, then put his hand in his pocket and pulled out a cigarette case. "You might like this as a souvenir —Torloni mislaid it. There's something interesting inside." He tossed it on the table and left the saloon.

I looked at Francesca and Coertze, then slowly put out my hand and picked it up. It had the heavy, familiar feel of gold, but I felt no sudden twist to my guts as I had when Walker had put the gold Hercules into my hand. I was sick of the sight of gold.

I opened the case and found a letter inside, folded in two. It was adressed to me, care of the yacht *Sanford,* Tangier Harbour, and had been opened. I started to read it and began to laugh uncontrollably.

Francesca and Coertze looked at me in astonishment. I tried to control my laughter but it kept bursting out hysterically. "We've . . . we've won . . . won a sweep . . . a lottery," I gasped, and passed the letter to Francesca, who also started to laugh.

Coertze said blankly, "What lottery?"

I said, "Don't you remember? You insisted on buying a lottery ticket in Tangier—you said it was for insurance. It won!"

He started to smile. "How much?"

"Six hundred thousand pesetas."

"What's that in money?"

I wiped my eyes. "A little over six thousand pounds. It won't cover expenses—what I've spent on this jaunt—but it'll help."

Coertze looked sheepish. "How much did you spend?"

I began to figure it out. I had lost *Sanford*—she had been worth about £12,000. I had covered all our expenses for nearly a year, and they had been high because we were supposed to be wealthy tourists; there had been the exorbitant rental of the Casa Saeta in Tangier; there was the outfitting and provisioning of the boat.

I said, "It must run to about seventeen or eighteen thousand."

His eyes twinkled and he put his hand to his fob pocket. "Will these help?" he asked, and rolled four large diamonds on to the table.

"Well, I'm damned," I said. "Where did you get those?"

"They seemed to stick to my fingers in the tunnel." He chuckled. "Just like that machine pistol stuck to yours."

Francesca started to giggle and put her hands to her breast. She produced a little wash-leather bag which was slung on a cord round her neck and emptied it. Two more diamonds joined those on the table and there were also four emeralds.

I looked at both of them and said, "You damned thieves; you ought to be ashamed of yourselves. The jewels were supposed to stay in Italy."

I grinned and produced my five diamonds and we all sat there laughing like maniacs.

IV

Later, when we had put the gems away safe from the prying eyes of Metcalfe, we went on deck and watched the hills of Spain emerge mistily from over the horizon. I put my arm round Francesca and said wryly, "Well, I've still got a half-share in a boatyard in Cape Town. Will you mind being a boat-builder's wife?"

She squeezed my hand. "I think I'll like South Africa."

I took the cigarette case from my pocket and opened it with one hand. The inscription was there and I read it for the first time—"*Caro Benito da parte di Adolf—Brennero—1940.*"

I said, "This is a pretty dangerous thing to have around. Some other Torloni might see it."

She shivered and said, "Get rid of it, Hal; please throw it away."

So I tossed it over the side and there was just one glint of gold in the green water and then it was gone for ever.

THE END

Alistair MacLean

His first book, *HMS Ulysses*, published in 1955, was outstandingly successful. It led the way to a string of best-selling novels which have established Alistair MacLean as the most popular thriller writer of our time.